AI LESSONS FOR THE CLASSROOM

A Practical Guide for Educators

By John C. Kim II

© Copyright by John C. Kim II, 2026.
All Rights Reserved.

No part of this publication may be reproduced, distributed, or transmitted
in any form without prior written permission.

Table of Contents

Introduction: The Answer Is 42 3
    What This Book Is 3
    The Teacher's Advantage 4
    How to Use This Book 5
    What AI Is (In Plain English) 5
    The Question Problem 6
    The Prompt as Pedagogy 6
    The Real Lesson 7
Chapter 1: The Ten-Minute Slide Deck 7
    The Use Case 8
    The Pedagogy 8
    The Method 9
    From the Field 11
    What Goes Wrong 12
    Going Further 13
    Your Toolkit 13
    The Real Work 14
    Example Prompts and Variations 15
Chapter 2: The AI Lab 17
    The Use Case 17
    The Pedagogy 17
    The Method 18
    From the Field 20
    What Goes Wrong 21
    Going Further 22
    Your Toolkit 23
    The Real Work 23
    Example Prompts and Variations 24
Chapter 3: The Fact-Check Game 26
    The Use Case 26
    The Pedagogy 26
    The Method 28
    From the Field 29
    What Goes Wrong 30
    Going Further 31
    Your Toolkit 32
    Example Prompts and Variations 33
Chapter 4: 20 Questions with a Machine 35
    The Use Case 35

The Pedagogy	35
The Method	36
From the Field	38
What Goes Wrong	39
Going Further	39
Your Toolkit	40
Example Prompts and Variations	41
Chapter 5: Name That Plant	**43**
The Use Case	43
The Pedagogy	43
The Method	44
From the Field	45
What Goes Wrong	47
Going Further	47
Your Toolkit	48
Example Prompts and Variations	50
Chapter 6: Guess the Object	**51**
The Use Case	51
The Pedagogy	51
The Method	52
From the Field	53
What Goes Wrong	55
Going Further	55
Your Toolkit	56
Example Prompts and Variations	58
Chapter 7: Argue With Me	**60**
The Use Case	60
The Pedagogy	60
The Method	62
From the Field	63
What Goes Wrong	64
Going Further	65
Your Toolkit	66
Example Prompts and Variations	67
Chapter 8: Group Chat	**69**
The Use Case	69
The Pedagogy	69
The Method	71
Sample Personas for the Council	72
From the Field	73

- What Goes Wrong 74
- Going Further 75
- Your Toolkit 75
- Example Prompts and Variations 77
- Chapter 9: Build a Model 79
  - The Use Case 79
  - The Pedagogy 79
  - From the Field 81
  - The Method 82
  - The PDSA Framework 83
  - What Goes Wrong 84
  - Going Further 85
  - Your Toolkit 85
  - Example Prompts and Variations 88
- Chapter 10: Quick Wins 89
  - The Research Foundation 90
  - A Note on Adaptation 96
- Chapter 11: When the Machine Breaks 96
  - The Pedagogy 97
  - The Method 98
  - Failure Type One: The Invented Source 99
  - Failure Type Two: The Confident Error 100
  - Failure Type Three: The Sycophant 101
  - Failure Type Four: The Math Mirage 103
  - Failure Type Five: The Bias Reflection 104
  - Going Further 107
  - Your Toolkit .... **Error! Bookmark not defined.**
  - Quick Reference: Failure Types 108
- CONCLUSION: THE QUESTION PROBLEM 110
- Appendix A: ADDITIONAL AI CLASSROOM USE CASES FOR STUDENTS 112
- Appendix B: General Purpose Prompts 118
- Appendix C: Assessment Rubrics for AI-Assisted Work 120
  - Implementation Guidance 128
- Appendix D: Failure Types at a Glance 128
- Appendix E: Tools and Platforms to Explore 130
- References 133

Introduction: The Answer Is 42

In 1979, Douglas Adams published *The Hitchhiker's Guide to the Galaxy*. In it, a race of hyper-intelligent beings builds a supercomputer called Deep Thought to answer the ultimate question: What is the meaning of life, the universe, and everything?

Deep Thought takes 7.5 million years to compute the answer.

The answer is 42.

When the beings express their dismay at this seemingly meaningless response, Deep Thought offers a diagnosis. "I think the problem, to be quite honest with you, is that you've never actually known what the question is."

Adams meant it as satire. He could not have known that forty-five years later, we would carry Deep Thought in our pockets.

Today, high school students can ask ChatGPT to explain quantum entanglement, help with a college essay, fix Python code, or summarize the themes of Beloved. The machine responds in seconds. It sounds confident and is often correct. When wrong, it is wrong with unwavering certainty.

We have solved the answer problem. Any answer, on any topic, is now available to anyone with a phone. The library of human knowledge has been compressed into a chat window.

We have the answers to nearly every question in the palm of our hands. It is not about knowing the answer anymore but knowing the right questions to ask.

The beings in Adams's novel spent seven and a half million years waiting for an answer they didn't understand because they had never clarified what they were asking. They built the most powerful computer in the universe and pointed it at a vague abstraction. The machine did exactly what they asked. The failure was theirs.

This is the situation in every classroom where AI has arrived, which is to say, every classroom. Students now have access to a machine that can produce fluent, plausible text on any subject. The machine will give them answers. Whether those answers hold significance is uncertain, as is the students' ability to recognize the difference.

AI should be used to surface thinking, not replace it, and the teacher's role is to design friction, not efficiency.

What This Book Is

This is not a manifesto or a policy document. It is not a breathless prediction about education's transformation or a dire warning about its collapse. It is a field guide built from my experience teaching with AI in middle school and high school.

Each chapter gives you one specific thing to try on Monday morning. Not a theory. A practice. Complete with the exact prompts you will type and the mistakes your students will make. Some chapters take ten minutes. Others unfold over a semester. All of them treat AI as what it actually is, a tool that amplifies whatever you bring to it.

If you bring vague questions, you will get plausible nonsense. If you bring sharp questions, you will get useful material to think with. The machine does not care. It has no preferences. It is a mirror that reflects the quality of what you put in front of it.

Your job, as a teacher, is to teach students how to stand in front of that mirror.

The Teacher's Advantage

You do not need to master AI. You do not need to understand neural networks, transformer architectures, or the mathematics of large language models. You need to understand one thing: the machine gives you what you ask for, and most people ask badly.

Your advantage is that you have spent years learning to ask nicely. You have listened to students struggle to articulate their thoughts. You have learned to rephrase questions when they are wrong. You have developed an instinct for when understanding is real and when it is performance. You have built relationships that no algorithm can replicate.

AI is powerful. It is also profoundly stupid in ways that matter. It does not know your students. It does not know what they struggled with yesterday or what they might become tomorrow. It cannot tell when a student's confidence is masking confusion. It cannot sense when frustration has tipped from productive to destructive. It cannot look a teenager in the eye and say, "I know this is hard. Let's try a different way."

Consider what you do every day that no AI can replicate. You notice when a student's confidence is masking confusion. You sense when frustration has tipped from productive to destructive. You remember that Marcus struggled with fractions last month, so you frame today's algebra problem differently for him. You see the flicker of understanding in Sarah's eyes and know she's ready for a more complex question. These skills include adaptive expertise, pedagogical content knowledge, and relational trust. AI has none of them. It cannot read a room. It cannot adjust in real time. It cannot look a teenager in the eye and say, 'I know this is hard. Let's try a different way.' The technology changes. The craft endures.

How to Use This Book
Each chapter follows the same structure:
- **The Use Case:** What you are trying to accomplish, in concrete terms.
- **The Pedagogy:** Why this matters for learning. What research or theory informs the approach.
- **The Method:** Step-by-step instructions. Exact prompts you can copy. Timing suggestions.
- **From the Field:** A story from a real classroom. What actually happened when a teacher tried this.
- **What Goes Wrong:** Common failures and how to avoid them. The traps other teachers fell into.
- **Going Further:** Extensions for ambitious teachers or advanced students.
- **Your Toolkit:** Specific tools and resources. What to download, what to bookmark.

You do not need to read the chapters in order. Start with whatever problem is most urgent in your classroom. If your students are submitting AI-generated essays, start with the fact-checking chapter. If they are disengaged, try the debate chapter. If you want to get outside, start with the nature identification exercises. This is a handbook, not a novel. Use what you need.

What AI Is (In Plain English)
Generative AI systems like ChatGPT, Claude, and Google Gemini are pattern-completion machines. They have consumed vast libraries of human writing and learned to predict what words come next. They do not think. They do not understand. They have no beliefs, no desires, no memory of your previous conversation once the window closes. They are remarkably good at producing plausible text, images, and code based on your instructions.

Think of them as talented impersonators with no opinions of their own. Ask for a Shakespearean sonnet, and you get something that sounds like Shakespeare. Ask for a business memo, and you get corporate prose. Ask for medical advice, and you get something that sounds like a doctor, which is dangerous precisely because it sounds so convincing.

The technical term for their failures is "hallucination." The machine will confidently cite nonexistent studies, invent historical events, and attribute quotes to people who never said them. It does this because it is not retrieving facts from a database. It is generating sequences of words that statistically resemble factual statements. Sometimes the resemblance is perfect. Sometimes it is a lie wearing truth's clothing.

This is not a flaw to be fixed. It is the nature of the technology. Understanding this changes how you use it.

The Question Problem

In the old model, answers were scarce. Teachers possessed what students needed. In the classroom, the teacher spoke, the students listened, and information flowed in one direction. Tests measured whether data had been received.

That model assumed scarcity. It made sense when the library was across town and closed on Sundays, when encyclopedias cost a month's salary, and when experts were inaccessible. Under those conditions, the teacher's job was to be the local repository of knowledge.

Those conditions no longer exist.

Today, a student with a phone has access to more information than any teacher could memorize in a lifetime. The scarcity has shifted. Answers are abundant. What is scarce is the ability to ask the right question, to evaluate whether an answer is trustworthy, to know when to press further and when to stop, and to synthesize information into understanding.

The best teachers have never just transmitted information. They have modeled curiosity. They have asked questions that opened doors. They have pushed students past easy answers into productive confusion. They have known which struggle was worth it and which was merely frustrating. None of this has changed. What has changed is that these skills, once a nice addition to content delivery, are now the entire point.

The Prompt as Pedagogy

There is a new skill in the world. It has an awkward name: prompt engineering. It means knowing how to talk to AI systems, so they give you functional responses.

The name makes it sound technical. It is not. Prompt engineering is the art of being specific. It is knowing what you want before you ask for it. It anticipates misunderstanding and heads it off. It provides context, sets constraints, and specifies the format. It is, in other words, everything a good teacher has always demanded of student writing.

"What's your thesis?" a writing teacher asks. "Who is your audience? What do they already know? What are you trying to make them believe? What evidence do you have? What would someone who disagrees say?"

These are the same questions that produce a good AI prompt. A student who can write a clear prompt can write a clear essay. A student who cannot specify what they want from a machine probably cannot identify what they think about a topic. The skills transfer because they are the same skills, clarity of thought expressed in precision of language.

This is the hidden opportunity. Every AI interaction is a writing exercise. Every prompt is a chance to practice specificity. Every failed response is feedback that you were not clear enough. Try again.

The Real Lesson

Douglas Adams chose the number 42 as a joke. "It was a joke," he later wote. "It had to be a number, an ordinary, smallish number, and I chose that one... I sat at my desk, stared into the garden, and thought, '42 will do.' I typed it out. End of story."

The joke landed because it pointed at something true. We want answers. We build machines to give us answers. And when the answers come, we discover that they were never the point.

Deep Thought understood this. After delivering its disappointing answer, it proposed building an even larger computer to determine what the question actually was. That computer was called Earth. Its program would run for ten million years, and its processors would be the living beings who inhabited it.

The search for meaning, in Adams's universe, was never something a machine could do alone. It required life. Curiosity. Argument. Mistake. Revision. All the messy, inefficient processes that make learning real.

Your classroom is one of those processors. The students in it are working, whether they know it or not, on questions that matter. AI can help them. It can also distract them, mislead them, or give them the illusion of understanding where none exists. The difference depends on you.

Teach them to ask better questions. The answers will take care of themselves.

Chapter 1: The Ten-Minute Slide Deck

Phase	Time Required	Description & Activities
Day 1: Generation & Start	45–50 Minutes	Intro (5m): Topic & AI tool setup. Creation (10–15m): Slide generation. Round 1 (25–30m): First 6–8 presenters.
Days 1–2: Round 1 (Cont.)	~100 Minutes Total	Completion of all first-round presentations (4 mins per student). Optional: Use Fishbowl model to save time.
Interim: Prep Time	20–30 Minutes	Students anticipate interrogation questions and prepare handwritten notes (Class time or Homework).
Days 3–4: Round 2	150–200 Minutes	Second presentations with extended questioning (5–8 mins per student). Usually spans 3–4 class periods.
Total Duration	2–4 Class Periods	

The Use Case

Students create a polished presentation on an assigned topic using AI-assisted design tools. The slides take ten minutes to generate. The learning takes much longer.

This chapter covers what happens after the slides exist. The AI handles the production. You handle the interrogation.

The Pedagogy

Presentation design has always been a bottleneck. Students spend hours choosing fonts, hunting for images, and adjusting layouts. They confuse aesthetic labor with intellectual labor. A student can spend an entire evening making a slide deck look professional without ever deeply engaging with the content on those slides.

AI tools like Canva's Magic Design, Gamma.app, and Tome collapse this timeline. A student can now generate a complete, visually coherent presentation in minutes. The slides will have appropriate images, consistent formatting, and logical structure. The AI handles the graphic design.

This sounds like cheating. It is not. It is a shift in where the work happens.

When production is cheap, evaluation becomes expensive. When anyone can generate a slide deck, the question becomes Does the presenter understand what is on those slides? Can they explain it? Can they defend it? Can they answer questions that go beyond what the AI provided?

This is where pedagogy lives. The presentation becomes a starting point, not an endpoint. The slides are a map of the territory the student must now explore on foot. Your job is to make them walk.

Bloom's taxonomy puts 'remember' and 'understand' at the base, 'analyze' and 'evaluate' at the top. AI excels at the base. It produces surface understanding. It cannot answer the follow-up question, explain why it made a choice, or connect content to last week's lesson.

When you probe a student's AI-generated presentation, you are climbing the taxonomy. You are forcing the student to move from "I can read this slide" to "I can explain what this means" to "I can apply this concept to a new situation." The AI gave them a scaffold. You make them climb it.

The Method
Phase One: Generation
1. Assign each student a specific topic within your unit. For an engineering class, this might be bridge types, renewable energy sources, manufacturing processes, or materials science concepts. The topic should be narrow enough to cover in six slides but rich enough to support deep questioning.
2. Provide access to an AI presentation tool. Gamma.app works well for beginners because it generates complete decks from a single prompt. Students type "Create a six-slide presentation about [topic] for a high school engineering class."
3. Set a time limit for generation, ten to fifteen minutes. The constraint matters. You are not asking students to perfect their slides. You are asking them to produce raw material quickly.
4. Students may edit the AI output, but they must preserve the core content for now. Changes come later, after they understand what they have.

Phase Two: First Presentation
5. Each student presents their slides to you. Set a strict time limit of two minutes maximum. Provide a rubric that includes clarity, organization, and visual design, but make clear that this is only the beginning.

6. Watch what happens. Most students will read directly from the slides. They will stumble over words they do not understand. They will rush through complex concepts without explanation. This is expected. This is data.
7. Do not interrupt during the two-minute presentation. Let them finish. The interrogation comes next.

Phase Three: The Interrogation

After the presentation, begin asking questions. This is where the learning happens. Your questions should move through several categories:

Category	Sample Questions
Vocabulary	"Define [term] in your own words." "What's the difference between X and Y?"
Restatement	"Explain this to a fifth-grader." "Say that without looking at your slides."
Verification	"Where did this statistic come from?" "How would you verify this?"
Application	"Give me a real-world example." "Where would you see this in our school?"
Connection	"How does this relate to last week?" "What would someone who disagreed say?"
Limitation	"What are the disadvantages?" "What did your presentation leave out?"

Do not ask every question for every student. Read the room. If a student clearly understands their material, push toward application and connection questions. If a student is struggling with basic vocabulary, stay there until the foundation is solid. The interrogation is diagnostic. It reveals what each student actually knows.

Phase Four: Preparation

After the first round, give students time to prepare for round two. They now know what kinds of questions you ask. They know that reading from slides will not be enough. Require them to prepare handwritten notes anticipating your questions. One to two pages minimum for a six-slide presentation. The notes must be in their own words, not copied from the AI output.

Phase Five: Second Presentation
The second presentation follows the same structure, but the dynamic has changed. Students now come prepared for interrogation. The total time per student expands from two minutes to five to eight minutes, a brief presentation followed by extended questioning. The slides remain the same, but the understanding behind them has deepened.

From the Field
I tried this with my engineering students last semester. Each student was assigned a topic: truss bridges, hydraulic systems, injection molding, and composite materials. They generated their slides in fifteen minutes using Gamma. The presentations looked professional. The content was accurate. Everything seemed fine.

Then came the first round of presentations.

The pattern was immediate and universal. Students read from their slides verbatim. Their eyes stayed fixed on the screen. When they encountered a technical term, they pronounced it carefully but without comprehension. One student presented on product improvement and discussed "Kaizen" without knowing what "Kaizen" meant. Another created a list of tools to use, but had no idea what any of them did.

The AI had given them words. It had not given them understanding. The students are very capable of phonological decoding (the mechanical process of turning letters into sounds), but do not have lexical competence or semantic depth (complete understanding of what they are saying or reading).

So I asked questions. "You just said 'Kaizan." What does Kaizan mean?" Silence. "Your slide provided a list of tools. Can you give me an example of how and where to use one of those tools?" Uncertainty. "What does the word exacerbate mean? Read the sentence and see if you can infer the meaning." A guess, then a better guess, then something approaching an answer.

The interrogation revealed the gaps. It also began to fill them. When a student could not define a word, we worked through it together. I asked whether the word sounded like one they had heard or whether they could replace it with another. The student arrived at the concept through questions, not through slides.

I told them there would be a second round. Same slides. Same rubric. Now they knew what was coming.

The transformation was remarkable. For their six-slide presentations, students arrived with one to two pages of handwritten notes. They had looked up the vocabulary they did not know. They had found real-world examples. They had anticipated my questions and prepared answers. One student who had stumbled over "iterative" in round one now explained it using the example within the engineering cycle. Another student had a list of examples for each definition.

The second-round presentations took four to six minutes each, a two-minute presentation followed by extended questioning. The questioning had changed character. I was no longer exposing ignorance. I was exploring understanding. The students could handle follow-up questions because they had done the work to understand their material.

The slides had not changed. The students had.

What I discovered was that the presentation was never the point. It was the starting point. The AI had given each student a map of unfamiliar territory. The first presentation revealed how little they knew about that territory. The interrogation forced them to explore it. The preparation for round two was the actual learning. And the second presentation was the demonstration that learning had occurred.

Future lessons only took a single round of presentations from then on.

I now think of the AI-generated slides as a vocabulary list for a language the students do not yet speak and concepts they do not know. The slides give them the words. My questions make them use those words until the words become theirs.

What Goes Wrong

- **Students accept AI content without review.** The AI will occasionally include inaccurate information, outdated statistics, or misleading simplifications. Students who do not verify the content will confidently present errors. Fix: Require a verification checklist. For each factual claim, students must identify a source that confirms it. If they cannot find confirmation, they must flag the claim as unverified.
- **The interrogation becomes adversarial.** If students feel attacked by your questions, they will shut down. The goal is not to embarrass them but to teach them. Fix: Frame the questioning as collaborative. "Let's figure this out together." "I'm asking because I want to understand your thinking." Celebrate when students admit they don't know something; that honesty is the beginning of learning.
- **Students prepare for the wrong questions.** Some students will memorize definitions without understanding them. They can recite what "thermal conductivity" means; however, they cannot explain why it matters. Fix: Vary your question types. Do not let students

predict exactly what you will ask. Application questions and connection questions resist memorization.
- **The time requirements spiral.** With 25 students and eight minutes each, you need more than three hours. Spread presentations across multiple days. Or use a fishbowl model, one student presents while others observe and take notes on the questioning techniques. Rotate who presents each day.
- **Students hide behind the AI's choices.** When asked why they organized their presentation a certain way, students say, "That's how the AI did it." They abdicate ownership. Fix: Require at least one structural change before round two. Students must add, remove, or reorder at least one slide and justify the decision. This forces them to take ownership of the content.

Going Further
- **The One-Slide TED Talk.** Students limit their presentation to a single slide that supports a two-minute verbal presentation. No bullet points allowed on the slide. This forces them to internalize the content completely; they cannot read from a single image. The constraint reveals who truly understands the material.
- **Peer Interrogation.** Train students to question each other using the same categories you use. Assign each student a "questioner" partner. The questioner's grade depends partly on the quality of their questions. This multiplies the interrogation and teaches students to think critically about others' presentations.
- **The Revision Challenge.** Students take their AI-generated deck and systematically improve it over three iterations. After each iteration, they write a brief reflection: What did I change? Why? What did I learn by making this change? The final submission includes all versions plus the reflections.
- **Cross-Topic Connections.** After all students have presented, assign each student to identify three connections between their topic and someone else's. The student who presented on composite materials must connect to the student who presented on manufacturing processes. This forces synthesis across the unit.
- **The Teaching Test.** Students must teach their topic to a younger student or to a family member who knows nothing about engineering. They record the teaching session or bring a signed note confirming it happened. If you can teach it, you understand it. If you cannot, you have more work to do.

Your Toolkit

AI Presentation Tools
- **Gamma.app:** Best for beginners. Generates complete presentations from a single prompt. Free tier available with limited exports.
- **Canva Magic Design:** Integrated into Canva's broader design platform. Suitable for students who want more control over visual elements.
- **Tome:** More sophisticated output, better for older students. Can generate narrative-style presentations.
- **Beautiful.ai:** Strong templates and formatting. Useful when visual consistency matters.
- **Microsoft Designer / PowerPoint Copilot:** For schools with Microsoft 365 subscriptions. Integrated into familiar tools.

Supporting Resources
- **Question stems document:** Create a one-page handout of question types for your own reference during interrogations.
- **Verification checklist template:** A simple form where students record each factual claim and its source.
- **Reflection prompts:** Pre-written questions for students to answer after each presentation round.
- **Rubric with interrogation component:** Modify your existing presentation rubric to include points for responding to questions, defining vocabulary, and demonstrating understanding beyond the slides.

The Real Work

There is a version of this exercise that feels like cheating. Students generate slides, read them aloud, and collect their grade. The AI did the work. The student performed it. Nothing was learned.

In most versions of this activity in my class, AI-generated slides are the least important part of the process. The class is not about formatting slides. The slides exist so that you have something to ask questions about. The questions exist to help students think. The thinking is the learning.

The difference between these two versions is you. The AI does not care which version you choose. It will produce the same slides regardless of the method. However, you care about this detail, and your students should, too.

When a student can explain their slides without reading them, answer questions they did not anticipate, and connect today's topic to last week's lesson and next month's project, they have learned something. The AI gave them a starting point. You made them earn the destination.

Example Prompts and Variations
Generating an Initial Outline
Purpose: *Produces raw material for students to evaluate, verify, and improve*
Create a 5-slide presentation outline about [TOPIC]. For each slide, provide a title and three bullet points. Include one statistic or fact per slide.
The Verification Challenge
Purpose: *Teaches students to fact-check AI output and builds healthy skepticism*
I'm going to show you a presentation about [TOPIC]. For each claim or statistic, tell me: (1) Is this accurate? (2) What's your confidence level? (3) What source would I need to verify this?
Asking for Alternative Perspectives
Purpose: *Reveals AI blind spots and encourages critical thinking about bias*
This presentation argues [MAIN POINT]. What perspectives or counterarguments are missing? Who might disagree with this framing, and why?
Improving Visual Storytelling
Purpose: *Develops presentation literacy and visual communication skills*
I have a slide with this content: [PASTE CONTENT]. Suggest three different ways I could visualize this information. Which approach would be most memorable for an audience, and why?
The One-Slide TED Talk
Purpose: *Forces the distillation and prioritization of key messages*
Help me condense [TOPIC] into a single powerful slide. What is the one essential idea, the one compelling image, and the one statistic that would make an audience remember this?

Chapter 2: The AI Lab

Phase	Time Required	Description & Activities
Class 1: Baseline Test	45–50 Minutes	Group Prep (20m): Develop questions. Testing (15m): Run AI systems. Analysis (10m): Compile initial results.
Class 2: The Breaking Point	45–50 Minutes	In-Class: Probe edge cases & document failures with hypotheses. Homework: 20–30m of additional exploration.
Class 3: Ethics Audit	45–50 Minutes	Framing (10m): Set context. Audit (25m): Structured investigation. Discussion (15m): Class-wide share out.
Class 4: Improvement Cycle	45–50 Minutes	Prompt Engineering: Select a previously failed task and systematically test new prompting strategies to improve output.
Class 5: The Report	90 Minutes (2 sessions)	Session A: Poster/Visual preparation. Session B: Presentations and audience Q&A.
Full Unit Totals	5–7 Hours Total	Spread across the unit for maximum depth of investigation.

The Use Case

Students become researchers. Instead of passively using AI tools, they systematically test them, document their failures, probe their boundaries, and develop critical frameworks for evaluating machine intelligence. The course unfolds over several weeks, transforming students from consumers of AI into investigators of it.

This is a thinking class that happens to use technology as its subject. Use a topic from your class for the students to investigate.

The Pedagogy

In 1999, Sugata Mitra conducted an experiment that would reshape how we think about learning and technology. He installed a computer in a hole in a wall separating his office in New Delhi from an adjacent slum. The children who lived there had never seen a computer. No one taught them how to use it.

Within hours, they were surfing the web. Within days, they were teaching each other. Within months, they had developed their own vocabulary for what they saw on screen, calling the hourglass cursor "damru" after an hourglass-shaped drum and the mouse pointer "sui," the Hindi word for needle.

Mitra called this "minimally invasive education." His findings, replicated across India and later in the UK, suggested that children could teach themselves and each other when given access, agency, and the right kind of challenge. The key was curiosity, not instruction.

The AI Lab applies this principle to generative AI. Instead of lecturing students about what ChatGPT can and cannot do, you give them the tools and a systematic framework for discovery. They become researchers investigating a new kind of intelligence. Their findings are genuinely uncertain. Their conclusions are their own.

This approach builds precisely the skills that the current moment demands. A 2016 Stanford University study found that more than 80 percent of middle school students could not distinguish between sponsored content and real news stories on a website. A follow-up study in 2019 found that 96 percent of high school students failed to consider why a climate change website's ties to the fossil fuel industry might affect its credibility. The researchers' conclusion was stark: young people's ability to reason about information online is "bleak."

AI makes an already dire media literacy crisis actively worse. Students who couldn't evaluate human sources now face machine sources that are more confident, more prolific, and harder to trace. We are not starting from zero. We are starting from negative. It does not signal uncertainty. It does not flag its own errors. If students cannot evaluate sources from humans, how will they assess outputs from machines?

The AI Lab teaches evaluation by making students evaluators. They do not read about AI limitations. They discover them. They do not memorize lists of AI failures. They create those lists themselves.

The Method

The AI Lab unfolds over five weeks. Each week has a distinct focus, but the activities build on each other. Students maintain a research journal throughout, documenting their findings, questions, and evolving understanding.

Week 1: The Baseline Test

Students create a standardized assessment to test AI capabilities. Working in small groups, they develop ten questions spanning multiple domains: factual recall, mathematical reasoning, logical puzzles, creative writing, local or specialized knowledge, ethical dilemmas, and tasks requiring real-time information.

Example questions might include "What is the population of our city according to the most recent census?" (local knowledge); "Write a haiku about autumn with exactly 5-7-5 syllables" (constrained creativity); "If it takes five machines 5 minutes to make five widgets, how long would it take 100 machines to make 100 widgets?" (logical reasoning); "Who won the most recent game played by [local sports team]?" (real-time information).

Each group runs their test on at least two different AI systems (ChatGPT, Claude, Gemini, or others). They score responses on accuracy, confidence level, and whether the AI acknowledged uncertainty. Results are compiled in a shared document.

Week 2: The Breaking Point

Assignment: "Find something this AI cannot do."

Students probe edge cases and document failures. The assignment requires more than cataloging errors. For each failure, students must develop a hypothesis about why it occurred. What does the failure reveal about how the system works?

One documented failure type involves syllable counting. Research has shown that large language models consistently struggle to count syllables accurately, even when they can perfectly describe what a syllable is. ChatGPT can explain the 5-7-5 structure of a haiku, then produce a haiku with a 7-6-5 structure while claiming it follows the traditional pattern. Students who discover this learn something fundamental: the AI is predicting plausible sequences of words, not reasoning about the properties of those words.

Other productive failure domains include mathematical word problems with unusual framing, tasks requiring spatial reasoning, questions about recent events, requests for information about less-documented communities or topics, and tasks requiring the AI to maintain consistent information across a long conversation.

Week 3: The Ethics Audit

Students test AI responses to ethically complex prompts and analyze the guardrails. What does the AI refuse to do? What does it do reluctantly? What does it do without hesitation that perhaps it should hesitate about?

This is an investigation of design choices. Why might the companies that built these systems have made certain decisions? What values are embedded in the refusals? Are those values consistent across different types of requests?

Students might discover that an AI will refuse to write a persuasive essay arguing for one political position but will readily explain the arguments on both sides. They might find that it handles some sensitive topics with extensive caveats while addressing others matter-of-factly. Each observation is data about human choices embedded in machine behavior.

Week 4: The Improvement Cycle

Students attempt to get better outputs through refined prompting. They select one task where the AI initially performed poorly and systematically test different approaches.

Strategies to test include providing more context, breaking complex tasks into steps, asking the AI to "think through" its reasoning before answering, providing examples of desired outputs, explicitly specifying format and constraints, and asking the AI to identify potential errors in its own response.

Students document which strategies work and which do not. The meta-lesson is that the quality of AI output depends heavily on the quality of human input. This is the introduction to prompt engineering as a skill, though the term need not be used.

Week 5: The Report

Student teams present their findings in a scientific poster session format. Each poster includes their research questions, methodology, key findings, hypotheses about underlying causes, and implications for how AI should (and should not) be used.

The session is open to other classes, administrators, or parents. Students field questions and defend their conclusions. The audience votes on the most significant finding and the most rigorous methodology.

From the Field

Documented classroom experiments with AI reveal consistent patterns in how students engage with these tools when given systematic frameworks for investigation.

This chapter highlights the tools I teach and my students use in my manufacturing and engineering classes. Engineers often test materials to failure, determining their breaking points. We frequently break items, especially those we construct. Stress tests are a common practice for welders. By using AI to document and research potential design failures, some of my students can avoid frustration when they invest significant time and effort into a project that may ultimately not meet specifications and fail to meet them.

Research published in *Frontiers in Education* found that the type of activity significantly influences how students perceive AI tools. When students engaged in structured coding activities in which they could verify AI outputs against expected results, they developed a more nuanced understanding of the technology's capabilities and limitations than students in open-ended discussion activities. The key variable was verifiability. When students could check whether the AI was right or wrong, they learned more about when to trust it.

A teacher at P.S. 54 in Staten Island, Joanna Stillman, described to NBC News how she approaches AI literacy with her students. She compares ChatGPT to "the drunk uncle" who "will give you information, but you don't know how true it is." In one lesson, she showed students three photographs and asked which was AI-generated. The students confidently identified which images they thought were real and fake. All three were AI-generated. "Their minds were blown," Stillman said.

This moment of revelation is the pedagogical core of the AI Lab. Students arrive with assumptions about AI. The structured investigation challenges those assumptions with evidence they generate themselves. The blown minds are the beginning of genuine critical thinking.

Research on ChatGPT's performance in specialized domains reinforces what students discover through their own testing. Studies have documented notable failure rates in areas requiring a nuanced understanding. ChatGPT can describe the principles of haiku with perfect accuracy, then produce haiku that violate those principles while claiming otherwise. It can explain what syllables are, but it cannot count them reliably. These are not random errors. They reveal something fundamental about how large language models work: they predict plausible text, not truthful text.

When students discover these patterns themselves, the lesson is more durable than any lecture could provide. They have seen the machine confidently claim something false. They will not forget.

What Goes Wrong

- **Students treat the exercise as a gotcha game.** Finding AI failures becomes entertainment rather than an investigation. The accumulation of errors leads to dismissiveness rather than understanding. Fix: Require analysis, not just documentation. For every failure, students must propose a hypothesis about why it occurred. What does this error reveal about the system? This shifts the focus from "look how stupid it is" to "let's understand how it works."
- **The novelty wears off.** After the first few discoveries, the exercise becomes repetitive. Students go through the motions. Fix: Introduce structured competition and increase the challenge. Week 1 focuses on apparent failures. Week 2 requires finding subtle shortcomings. Week 3 asks students to find cases where the AI is partially correct, which is harder to analyze than clear errors. Each week should feel more demanding than the last.

- **Students conclude that AI is useless.** A focus on failures can swing students too far in the opposite direction, from uncritical trust to blanket dismissal. Fix: Balance the investigation. Include explicit exploration of what AI does remarkably well. Students should emerge with calibrated expectations, not cynicism.
- **Students share inappropriate content.** The ethics audit week, in particular, can lead students to test problematic prompts. Fix: Establish clear boundaries before Week 3. Students are investigating design choices and are not trying to produce harmful content. Review and approve research protocols before students begin.
- **Findings are not generalizable.** Students test specific questions and assume results apply universally. An AI that fails one math problem might succeed at another. Fix: Require replication. If a student finds a failure, can they reproduce it? Can they identify similar prompts that produce the same failure? This is the scientific method in action.

Going Further
- **The Cross-Cultural Test.** Students test AI's knowledge of their own cultural backgrounds, languages, and community histories. Where does the AI's training data fail them? This investigation often reveals systematic gaps; the AI knows more about some communities than others. Students document whose knowledge is represented and whose is missing. This connects AI literacy to questions of representation and power.
- **The Time Capsule.** Students document current AI capabilities with detailed predictions for one year later. What do they expect to improve? What do they think will remain difficult? Seal the predictions. Open the capsule at the end of the following school year. Compare forecasts to reality. This builds awareness that AI is a moving target, not a fixed technology.
- **The Comparison Study.** Students run identical tests across multiple AI systems and create a detailed capability comparison. Which system is best for what? Are any consistently better across all domains? The exercise reveals that "AI" is not monolithic; different systems have different strengths.
- **The Bias Investigation.** Students systematically test for patterns that might indicate bias in AI outputs. They vary names, genders, locations, or other demographic markers in otherwise identical prompts. Do the responses differ? This is advanced work that

requires careful methodology, but it connects AI literacy to larger questions of fairness and equity.
- **The Publication.** Students compile their most significant findings into a report or website and share them beyond the school. This could be presented to the school board, published in a school newspaper, or posted online. The audience raises the stakes and gives the work purpose beyond the classroom.

Your Toolkit

AI Systems to Test
- **ChatGPT (OpenAI):** The most widely known system. Free tier available. Good baseline for comparison.
- **Claude (Anthropic):** Often gives different responses than ChatGPT. Useful for comparison studies.
- **Google Gemini:** Integrated with Google search. Interesting for testing real-time information access.
- **Perplexity:** Provides sources for claims. Useful for teaching source verification.

Documentation Tools
- **Shared Google Doc or Notion page:** For compiling class findings across groups.
- **Screenshot tools:** For documenting exact prompts and responses.
- **Research Journal template:** Structured format for individual student observations.

Assessment Resources
- **Scientific poster templates:** For Week 5 presentations.
- **Peer review rubrics:** For evaluating each group's methodology and conclusions.
- **Hypothesis testing framework:** Simple template for structuring claims about why AI behaves as it does.

The Real Work

Sugata Mitra's hole-in-the-wall experiment worked because children were given a genuine challenge with no predetermined answer. They were not following a script. They were exploring unknown territory. The discoveries were their own.

The AI Lab works for the same reason. Students are not memorizing facts about artificial intelligence. They are generating those facts through their own investigation. They are building the mental frameworks they will need to navigate a world where AI is everywhere and perfect trust is nowhere.

The children in New Delhi invented their own vocabulary for what they saw on screen. Your students will invent their own vocabulary for AI's failures and capabilities. They will develop their own rules of thumb for when to trust and when to verify. Those self-generated frameworks will last longer than anything you could have told them.

The role of the teacher in this model is not to provide answers. It is to provide structure, challenge, and the occasional seed for discovery. You are the granny who says, "Did you know the computer can play music?" and then steps back. The learning happens in the stepping back.

Example Prompts and Variations
Establishing a Baseline
Purpose: *Creates measurable data for comparing AI performance across domains*

I'm going to test your knowledge of [SUBJECT AREA]. Answer these questions to the best of your ability: [LIST 5-10 QUESTIONS]. For each answer, rate your confidence from 1-10.

Testing Knowledge Boundaries
Purpose: *Reveals temporal limitations and teaches students about training data*

What is the most recent event you know about regarding [TOPIC]? What is your knowledge cutoff date? What might have changed since then?

Probing for Hallucinations
Purpose: *Demonstrates how AI can confidently generate false information*

Tell me about [FICTIONAL OR OBSCURE TOPIC]. If you're not certain this exists, say so. If you're making assumptions, identify them.

Comparing Reasoning Approaches
Purpose: *Reveals AI reasoning processes and their potential failure points*

Solve this problem two different ways: [PROBLEM]. Show your reasoning step by step for each approach. Which method is more reliable, and why?

The Limits Test
Purpose: *Encourages AI self-awareness and teaches appropriate use cases*

I want to understand what you cannot do. Give me five examples of questions or tasks where you would likely fail or produce unreliable results.

Ethics Investigation
Purpose: *Opens discussion of AI ethics and responsible use*

I'm going to ask you to [POTENTIALLY PROBLEMATIC REQUEST]. Before you respond, explain: What ethical concerns might this raise? What safeguards do you have? What should I consider before using your response?

Chapter 3: The Fact-Check Game

Phase	Time Required	Description & Activities
Preparation	5–10 Minutes	Teacher Prep: Generate AI statements on the lesson topic and prepare the scoring sheet/rubric (done before class).
Class 1: Setup	5 Minutes	Launch: Display AI output, explain the categories (True, False, Misleading), and assign student teams.
Class 1: Verification	15–20 Minutes	The Hunt: Teams use lateral reading and search strategies to classify statements under a visible timer.
Class 1: Debrief	15–20 Minutes	Review: Class-wide scoring. Discuss "Hardest to Verify" statements and successful search techniques.
Total Core Time	45 Minutes	Standard class period duration.
Extended Option	60–75 Minutes	Add a deep-dive discussion on source credibility and the "SIFT" method for evaluation.

The Use Case

Students verify AI-generated claims using triangulated sources. The AI produces statements that may be true, false, or misleading. Students compete to classify each statement and cite their evidence correctly. The game builds media literacy, research skills, and a healthy skepticism toward confident-sounding text.

This lesson on verification uses AI as the source material.

The Pedagogy

AI hallucinations are not bugs. They are teaching opportunities.

When ChatGPT confidently states that a fictional battle occurred, invents a scientist who never existed, or cites a research paper that was never published, students confront the fundamental problem that plausibility in the information age is not truth. A statement can sound authoritative, be formatted appropriately, and include all the markers of credibility while still being completely false.

The research on AI hallucinations is sobering. A study published in *Scientific Reports* found that 55% of citations generated by GPT-3.5 were entirely fabricated, and 18% of citations from GPT-4 were fabricated. These fake citations included real author names, properly formatted digital object identifiers, and plausible journal titles. They looked exactly like real citations. They just happened to reference papers that did not exist.

The consequences of trusting such fabrications can be severe. In May 2023, a New York attorney named Steven Schwartz submitted a legal brief in *Mata v. Avianca* that cited six court cases generated by ChatGPT. The cases did not exist. The judges named as authors had never written the opinions attributed to them. When Schwartz asked ChatGPT to verify the cases, the AI assured him that they were real and could be found in legal databases such as Westlaw and LexisNexis. They could not. The court imposed a $5,000 sanction. Judge P. Kevin Castel noted that the AI had done "exactly what" the lawyer asked it to provide, cases supporting his argument. The problem was that the cases were fiction.

Students face the same challenge, often with lower stakes but similar dynamics. A 2024 study at the University of Mississippi found that 47% of student-submitted citations from AI sources had incorrect titles, dates, authors, or combinations of all three. The students were not being dishonest. They did not know that the AI was making things up.

Deep Thought took 7.5 million years to compute an answer. It never occurred to the machine to check whether the answer was true. It had no mechanism for verification, no concept of evidence, no way to distinguish between a response that corresponded to reality and one that merely sounded plausible. When the beings who built it expressed dismay at the answer "42," Deep Thought could not help them. It had done exactly what they asked. The failure was in the asking.

ChatGPT takes three seconds. It, too, has no mechanism for verification. It generates text that matches patterns in its training data. Sometimes those patterns correspond to facts. Sometimes they correspond to plausible-sounding fictions. The machine cannot tell the difference. That task falls to the humans who use it.

The Fact-Check Game teaches students to do what Deep Thought and ChatGPT cannot. It encourages them to ask whether an answer is true, seek evidence beyond the machine's confident assertions, and triangulate claims against independent sources. The beings in Adams's novel waited 7.5 million years for an answer they could not evaluate. Your students will learn to assess answers in minutes. That is the skill the age of AI demands.

This approach draws on research from the Stanford History Education Group, which has pioneered the concept of "lateral reading." Professional fact-checkers, the researchers found, do not evaluate a source by reading it closely. They leave the source and search for what other sources say about it. They triangulate. Students taught this method significantly outperform those given traditional source-evaluation checklists. The Fact-Check Game applies lateral reading to AI output.

The Method

Preparation (Before Class)

Prompt an AI to generate ten statements on a topic relevant to your current unit. For a history class, you might ask, "Give me ten facts about the French Revolution." For a science class, "Tell me ten things about how vaccines work." For an English class, "Provide ten facts about Shakespeare's life."

Do not edit the output. Do not screen for accuracy. The game works best when you do not know which statements are true, false, or misleading. You will verify alongside your students.

Prepare a scoring sheet with three columns for each statement: TRUE, FALSE, MISLEADING. Include space for students to note their sources.

The Game (In Class)

1. **Display the AI output.** Project the ten statements without commentary. Tell students, "AI generated these statements. Some may be true. Some may be false. Some may be technically true but misleading. Your job is to figure out which is which."
2. **Divide into teams.** Groups of three or four work well. Each team needs at least one device for research.
3. **Set the timer.** Fifteen to twenty minutes is usually sufficient for ten statements. The time pressure matters. It prevents endless rabbit holes and forces prioritization.
4. **Teams classify each statement.** For every classification, teams must cite at least two independent sources. Wikipedia can be one source, but not the only source. Fact-checking sites (Snopes, PolitiFact) count. Primary sources (academic papers, government data, original interviews) count double.
5. **Score the results.** Correct classification with no sourcing 0 points. Correct classification with a single source is worth 1 point. Correct classification with two or more sources: 3 points. Catching a MISLEADING statement (technically accurate but deceptive) 5 points. This scoring emphasizes that being right is not enough; you must show your evidence.
6. **Debrief.** Review each statement as a class. Which were hardest to verify? Which seemed true but were not? Which were technically accurate but missing important context? What search strategies worked best?

The Three Categories
TRUE: The statement is accurate and complete. Example: "Marie Curie won Nobel Prizes in both physics and chemistry."

FALSE: The statement is factually incorrect. Example: "The French Revolution began in 1799." (It began in 1789.)

MISLEADING: The statement is technically true but leaves out important context or implies something false. Example: "Vaccines contain formaldehyde." (True, but the amount is far less than what naturally occurs in a pear.)

The MISLEADING category is the most important and the most difficult. It teaches students that truth is not binary. A statement can be literally accurate while still being deceptive. This is where sophisticated media literacy lives.

From the Field
The problem this game addresses is well-documented across multiple research studies and high-profile cases.

Research published in *Mind Pad*, a journal of the Canadian Psychological Association, found that ChatGPT's "false citation rates" across psychology subfields ranged from 6% to 60%. The researcher, Jordan MacDonald, a PhD student at the University of New Brunswick, initially set out to identify ways to detect ChatGPT usage in student work. Instead, he discovered that "a lot of the references that ChatGPT cited did not actually exist." The fabricated citations were particularly insidious because they "often contain real authors, journals, proper issue/volume numbers that match up with the date of publication, and DOIs that appear legitimate."

This finding matters for students because it reveals a specific pattern in how AI fails. The AI is not generating random nonsense. It is generating plausible-looking nonsense. It has learned what citations are supposed to look like, so it produces text that matches that pattern. The content may be invented, but the form is correct. This is precisely the kind of error that a glance will not catch.

Duke University librarians have documented the same phenomenon and offer practical guidance for verification. They recommend searching for article titles with quotes in Google Scholar, checking journal titles against actual volume and page numbers, and searching for author publication lists independently. These are the exact skills the Fact-Check Game teaches.

In my engineering class, I recently had students use an AI tool to generate a curated list of problem-solving methodologies. Each student then selected one of these tools to research and explore in-depth. However, I quickly noticed that the AI overlooked that various industries often employ distinct approaches to problem-solving, leading to some ambiguity.

As the students made their selections, there was a noticeable element of "luck of the draw," as not every student ended up with a tool ideally suited to their specific context or industry. Furthermore, the information provided by the AI, while generally coherent, sometimes lacked relevance or specificity to the particular case studies we were discussing. In these situations, the AI was not being dishonest or fabricating details; rather, it simply lacked sufficient data to provide more accurate or tailored recommendations.

This experience prompted deeper discussion among the students regarding the intricacies of crafting effective prompts and the importance of asking the right questions. We delved into how refining our queries can lead to more relevant and applicable results, ultimately enhancing our problem-solving capabilities. This conversation highlighted the limitations of AI and offered an invaluable lesson in critical thinking and the art of inquiry in engineering and beyond.

The lesson for students is that you cannot verify AI claims by asking the AI. Verification requires leaving the source and checking independently. This is the principle of lateral reading, and it applies whether the source is an AI, a website, or a person.

What Goes Wrong

- **Students Google the exact AI phrase and accept the first result.** If the AI says, "The Battle of Thermopylae occurred in 480 BCE," students search that same phrase and find confirmation. They have not verified the claim; they have discovered text that contains the exact words. Fix: Require triangulation. No claim is verified until three independent sources confirm it. The sources must be genuinely independent, not copies of each other.
- **The game becomes rote.** After a few rounds, students develop a mechanical approach to search, find two sources, and move on. The critical thinking diminishes. Fix: Vary the domains. One week of historical facts. The following week, scientific claims. Next current events. Also, vary the difficulty; sometimes include statements that are almost entirely accurate with one small error. These require closer reading.
- **Students trust fact-check sites unquestioningly.** Snopes says it's false, so it's untrue. Fact-checking sites can be wrong, outdated, or

address a slightly different claim. Fix: Include a round where students must verify the fact-checkers themselves. What sources did Snopes use? Are those sources still accessible? Has new information emerged since the fact-check was published?
- **Students cannot find sources for factual statements.** Some accurate claims are complex to verify because they are obscure or specialized. Students mark them false because they cannot find evidence. Fix: Distinguish between "this is false" and "I could not verify this." Introduce a fourth category, UNVERIFIED. Teach that absence of evidence is not evidence of absence.
- **Time pressure leads to sloppy work.** Teams rush through statements to maximize their score, sacrificing accuracy for speed. Fix: Penalize wrong answers. A correct classification earns points; an incorrect classification loses points. This changes the incentive from "classify as many as possible" to "only classify what you are confident about."

Going Further
- **The Deep Fake Hunt.** Extend the game to AI-generated images. Students must determine whether an image is authentic, AI-generated, or manipulated. For each image, they must find the source or prove no original exists. Reverse image search becomes a key skill. This connects to broader concerns about synthetic media and visual misinformation.
- **The Citation Chain.** Students trace a claim backward through its sources. The AI says something. The AI's claim matches a Wikipedia article. The Wikipedia article cites a news story. The news story cites a research paper. Students follow the chain to find the source. How many layers until they reach primary evidence? What gets lost or distorted at each step? This teaches source literacy beyond the immediate claim.
- **The Confidence Calibration.** Before revealing the correct answers, ask students to rate their confidence in each classification on a scale of 1 (guessing) to 5 (certain). After scoring, compare confidence levels with accuracy. Are students well-calibrated? Do they know what they know? This builds metacognitive awareness about the difference between feeling certain and being correct.
- **Student-Generated Prompts.** Instead of the teacher generating AI statements, student teams create prompts on topics they choose. They generate ten statements, then swap with another team. Each team fact-checks the other's output. This increases ownership and

reveals which topics produce more hallucinations.
- **The Misinformation Autopsy.** When students find a false statement, they do not just mark it false. They investigate why the AI might have produced it. Was the AI confusing two similar events? Combining real facts in a false configuration? Making up details to fill gaps in its knowledge? This deeper analysis builds understanding of how misinformation is generated, not just how to detect it.

Your Toolkit

Fact-Checking Resources
- **Snopes (snopes.com):** One of the oldest fact-checking sites. Strong on viral claims, urban legends, and internet rumors.
- **PolitiFact (politifact.com):** Focuses on political claims. Useful "Truth-O-Meter" scale from True to Pants on Fire.
- **FactCheck.org:** Nonpartisan, from the Annenberg Public Policy Center. Strong on political and policy claims.
- **Google Fact Check Explorer:** Aggregates fact-checks from multiple sources. Useful for searching whether a claim has already been checked.
- **Wikipedia (for sourcing, not as a source):** The footnotes and references section of Wikipedia articles can point to primary sources. Teach students to use Wikipedia as a starting point, not a destination.

Image Verification Tools
- **Google Reverse Image Search:** Upload or paste an image URL to find other instances of the image online.
- **TinEye (tineye.com):** Specialized reverse image search. Shows where an image first appeared online.
- **InVID/WeVerify plugin:** Browser extension for verifying images and videos. Designed for journalists but accessible to students.

Media Literacy Games (for additional practice)
- **Bad News Game (getbadnews.com):** Players take the role of a misinformation creator. Research from Cambridge found it builds "psychological resistance" to fake news.
- **Reality Check (MediaSmarts):** Players evaluate social media posts and decide how to respond.
- **NewsFeed Defenders (FactCheck.org):** Quiz-based game teaching verification skills.
- **Checkology (News Literacy Project):** Comprehensive platform with lessons and assessments on news literacy.

The Real Work

The lawyer who submitted fake cases to a federal court was not stupid or dishonest. He was a professional with over thirty years of experience who trusted a tool he did not understand. When the AI told him the cases were real, he believed it. When doubts arose, he asked the AI to verify, and the AI reassured him. The system was designed to be helpful and confident, and it was both, right up until the moment it was caught.

Your students will face versions of this situation throughout their lives. They will encounter confident claims from machines, from websites, from people who seem to know what they are talking about. The matter at hand is whether they will verify or trust.

The Fact-Check Game does not teach students that AI is unreliable. It teaches them that everything requires verification. The AI is just the most convenient sparring partner, because it produces plausible-sounding claims at the push of a button, and some of those claims are wrong in ways that are hard to detect without effort.

The effort is the point. Skepticism is the skill. And the habit of checking, once formed, extends far beyond the classroom.

Example Prompts and Variations
Generating Mixed-Accuracy Statements
Purpose: *Creates raw material for the 'AI Says...' verification game*

Generate 10 statements about [TOPIC]. Make some entirely accurate, some partially accurate, some misleading, and some false. Do not tell me which is which. I will verify them myself.

The Source Challenge
Purpose: *Teaches research methodology and source hierarchy*

You said: '[AI STATEMENT]'. What sources would I need to verify this? Be specific about what type of source (primary, secondary, expert, official) and where I might find it.

Triangulation Practice
Purpose: *Develops triangulation skills and source independence awareness*

I found this claim: '[CLAIM]'. I've checked one source that says it's true. What two additional, independent sources should I check? What would make me confident that this is verified?

Spotting Misleading Framing
Purpose: *Teaches the difference between factual accuracy and truthful communication*

This statement is technically true: '[STATEMENT]'. How might it be misleading? What context is missing? How could someone use accurate facts to create a false impression?

Creating Fact-Check Cards
Purpose: *Generates structured materials for classroom fact-checking activities*

Create a fact-check practice card about [TOPIC]. Include: (1) A claim to verify, (2) Three potential sources to check, (3) What a verified answer would look like, (4) Common misconceptions to avoid.

Chapter 4: 20 Questions with a Machine

Phase	Time Required	Description & Activities
Class 1: Version 1	15–20 Minutes	AI Asks, Students Answer: Choose targets (2m), set up gameplay (10–12m), and analyze the AI's question sequence (5m).
Class 1: Version 2	20–25 Minutes	Students Ask, AI Answers: Strategy planning (5m), gameplay with justification (12–15m), and scoring comparison (5m).
Class 1: Version 3	15–20 Minutes	The Clue Game: High-speed variation focused on constructing efficient clues rather than multi-step questions.
Full Lesson Option	45–50 Minutes	Play two versions back-to-back. Includes a debrief on "efficient questioning" and strategy shifts.
Quick Version	10 Minutes	Brain Break: A single round used as a transition activity focused on curriculum-specific vocabulary.
Total Unit Time	30–45+ Minutes	Varies based on whether you run a single game or a whole class tournament.

The Use Case

Students play variations of the classic guessing game with AI. In one version, the AI asks questions and students answer. In another, students ask questions and the AI answers. In both versions, students learn to formulate precise questions, think systematically about categories, and observe how pattern recognition differs from understanding.

This is a game about thinking, not winning

The Pedagogy

In 1948, Claude Shannon published "A Mathematical Theory of Communication" and invented the field of information theory. Among his insights was a deceptively simple idea: information can be measured in bits, and a bit is the answer to a yes-or-no question.

Shannon showed that with 20 perfectly chosen yes-or-no questions, you can distinguish among over one million possibilities. Each question, if optimal, cuts the remaining options in half. The math is straightforward: $2^{20} = 1,048,576$. This is why the game of 20 Questions works. If you ask good questions, you can find any entity in a vast space of possibilities.

What makes the game pedagogically powerful is that not all questions are created equal. For example, "Is it bigger than a breadbox?" is more useful than "Is it an aardvark?" because it eliminates more possibilities.

The first question, well chosen, cuts the space roughly in half. The second question, poorly chosen, eliminates exactly one thing.

This reflects the essence of good thinking, which involves asking questions that maximize information gain. It is what scientists do when designing experiments, what doctors do when diagnosing, and what engineers do when troubleshooting. The skill transfers far beyond games.

The game Akinator demonstrates this principle at scale. Created by the French company Elokence, Akinator has been guessing characters since 2007. The game asks a series of questions and typically identifies who you are thinking of within 20 guesses, drawing from a database of millions of characters. The algorithm resembles a binary decision tree with probability weighting, in which each question dramatically narrows the possibilities. It learns from every game played, adding new characters when it fails and refining its questions based on accumulated data.

Modern AI chatbots can play both roles in this game. They can ask questions to guess what you are thinking, or they can answer questions about something they have "chosen." In either role, they reveal something important about how intelligence works, both in machines and in humans.

When the AI asks questions, students observe systematic categorization in action. The AI typically starts broad ("Is it living?") and narrows progressively ("Is it a mammal?" "Does it live in water?"). This demonstrates the power of hierarchical thinking.

When students ask questions, they must develop their own systematic approach. They cannot rely on random guessing. They must think about categories, eliminate possibilities efficiently, and formulate questions precisely enough for the AI to answer accurately. This is where the learning lives.

The pedagogical tradition here runs deep. Socrates himself used systematic questioning to help students discover knowledge they did not know they had. The "Socratic method" involves asking probing questions that encourage learners to examine their assumptions and develop their own understanding. Research shows that Socratic questioning allows assumptions to be tested and makes the thinking processes of critical analysis visible to students. The AI version of 20 Questions is a Socratic dialogue with a machine.

The Method

Version 1: AI Asks, Students, Answer

1. **Choose a topic domain.** For a science class, students might think of elements, animals, or scientific principles. For history, they

might think of historical figures, events, or documents, for literature, characters, authors, or literary devices. Constraining the domain makes the game more instructive and prevents students from choosing extremely obscure answers.

2. **Set up the AI prompt.** Tell the AI, "We are playing 20 Questions. I am thinking of a [domain]. Ask me yes-or-no questions to figure out what it is. After each of my answers, explain why you asked that question and what you learned from my answer."
3. **Students choose a target.** Each student (or pair of students) selects something within the domain. They write it down before the game begins. This prevents shifting answers.
4. **Play the game.** Students answer the AI's questions honestly. After each question, before moving on, discuss as a class: Was that a good question? What did the AI learn? What could the AI have asked instead?
5. **Track the question sequence.** Students record each question and answer. Afterward, they analyze which questions were most informative. Where did the AI go wrong? What category mistakes did it make?

Version 2: Students Ask, AI Answers (Reverse 20 Questions)

6. **Set up the AI prompt.** Tell the AI, "We are playing 20 Questions. You are thinking of a [domain]. I will ask yes or no questions to figure out what it is. Answer only yes or no. If my question cannot be answered with yes or no, tell me to rephrase it."
7. **Students plan their strategy.** Before asking any questions, students (individually or in teams) write down their first five planned questions. This prevents reactive, unstructured guessing.
8. **Play the game with a question budget.** Students have exactly 20 questions. They must track how many they have used. This creates strategic pressure to waste questions early, and you may not have enough to narrow down at the end.
9. **Require question justification.** Before asking each question, students must explain, "I am asking this because it will help me distinguish between X and Y." This makes the reasoning explicit.
10. **Score and compare.** Teams that answer fewer questions correctly score higher. Compare strategies. Which team asked the most efficient questions? Which had the best opening sequence?

Version 3: The Clue Game (AI Feeds Clues)

Instead of students answering questions, they provide the clues. Students give the AI one clue at a time about something they are thinking of. The AI tries to guess after each clue. Students track how many clues it takes for the AI to guess correctly.

The pedagogical value here is in clue construction. What information should you provide first? A broad category clue ("It's a type of animal") or a distinctive feature ("It's black and white")? How do you describe something without naming it directly? This is the inverse skill: asking good questions and providing good answers.

From the Field

The connection between 20 Questions and information theory has been used as a teaching tool for decades. CS Unplugged, a project that teaches computer science concepts without computers, includes a "Twenty Guesses" activity specifically designed to demonstrate Claude Shannon's insights about information and entropy. The activity shows students that the number of yes-or-no questions needed to identify something corresponds directly to the amount of information required to represent it.

I have adapted the 20 Questions game for a Biology and Botany class. One of the things I always struggled with in undergraduate class was species identification. It was long and tedious work. Asking hierarchical questions is a perfect fit for plant or animal identification. AI will even expand on more advanced subjects.

This principle underlies how Akinator achieves its seemingly magical results. Analysis of the game's algorithm suggests it uses a combination of binary search, decision trees, machine learning, and collaborative filtering. Through deductive reasoning, the system whittles millions of potential answers into a smaller subset of educated guesses. Each question is selected to divide the remaining possibilities as evenly as possible.

The educational value extends beyond computer science. Research in the field of inquiry-based learning has demonstrated that questioning skills are among the most critical cognitive competencies students can develop. A study on Socratic questioning in science inquiry-based learning found that systematic questioning helps students develop critical thinking skills and makes the reasoning process visible to both students and teachers. Students who learn to ask good questions, the research suggests, become better at reasoning in general.

Teachers have long recognized this. The Socratic method, despite its ancient origins, remains a cornerstone of modern pedagogy. Research on questioning in education shows that the level of questions asked influences the level of thinking that occurs. Higher-order questions (those that require analysis, synthesis, or evaluation) produce higher-order thinking. The 20 Questions game naturally pushes students toward higher-order questions because those are the questions that work.

What AI adds to this tradition is a patient, infinitely available sparring partner. The AI never gets bored. It never judges poor questions. It responds consistently, allowing students to run multiple experiments with different questioning strategies and compare results. This kind of systematic practice was difficult before AI made it possible.

What Goes Wrong

- **Students ask random or overly specific questions.** "Is it blue?" "Is it Abraham Lincoln?" These questions waste the limited budget. Fix: Require students to explain the expected information gain before asking each question. "If the answer is yes, I will have narrowed it to these possibilities. If no, I will have narrowed it to these." Make the reasoning visible.
- **The AI gives inconsistent answers.** The AI says something is "not living" and then says it "can grow." This happens because the AI is not actually reasoning about a specific object; it is generating plausible responses. Fix: When inconsistencies appear, stop and discuss. This is a teaching moment about how AI works; it predicts likely text, not truthful answers to logical questions. Have students deliberately try to catch inconsistencies.
- **Students treat it as pure entertainment.** The game is fun, which is good and students may lose focus on learning. Fix: Introduce the scoring and analysis phases. Competition and reflection transform play into practice. Require a written post-game analysis. What did you learn about effective questioning? What would you do differently next time?
- **Questions cannot be answered yes or no.** Students ask, "What color is it?" or "Where is it found?" These break the game format. Fix: Use a strict rule; if a question is not yes-or-no answerable, it counts against the 20-question limit but provides no information. This natural consequence trains precise question formation.
- **The domain is too broad or too narrow.** If students can think of anything in the world, the game becomes nearly impossible. If the domain is too narrow ("U.S. presidents"), the game is trivial. Fix: Calibrate the domain to the educational context. For a biology unit on classification, use "animals." For a chemistry unit, use "elements." The constraint should be large enough to challenge but small enough to succeed within 20 questions.

Going Further

- **The Information Theory Connection.** Introduce Claude

Shannon's concept of information entropy. Each question should provide approximately one "bit" of information, which means it should divide the remaining possibilities roughly in half. Students can calculate if there are 32 possible answers, how many perfect questions would you need? (5, because $2^5 = 32$.) This connects the game to mathematical foundations.
- **The Question Quality Rating.** After a round, have students rate each question on a scale of 1 to 5 for information value. What made some questions better than others? Can students articulate principles for good questioning that extend beyond the game?
- **The Category Tree.** Before playing, students create a hierarchical category tree for the domain. For "animals," they might start with "vertebrate/invertebrate," then branch into "mammal/bird/reptile/fish/amphibian" for vertebrates, and so on. The act of building the tree is itself a learning exercise in classification and organization.
- **Human vs. AI Questioning.** Students play the game twice, once with an AI that asks questions and once with a classmate. Compare who asked better questions? Where did the strategies differ? This reveals the difference between algorithmic and intuitive approaches to narrowing possibilities.
- **The Diagnostic Application.** Connect the game to real-world diagnostic reasoning. Show students how doctors use differential diagnosis, essentially a structured 20 Questions approach to identifying illness. How do engineers troubleshoot systems? How do scientists narrow hypotheses? The game becomes a model for professional reasoning.

Your Toolkit

AI Platforms for 20 Questions
- **ChatGPT, Claude, Gemini:** Any general-purpose AI can play 20 Questions with appropriate prompting. The advantage is flexibility; you can customize the domain and rules.
- **Akinator (akinator.com):** A specialized 20 Questions game for guessing characters. Helpful in demonstrating how well-designed algorithms can perform at this task. The AI asks questions; the user answers.
- **Custom GPTs:** Several purpose-built versions of ChatGPT exist specifically for playing 20 Questions. Search for "20 Questions" in the GPT store for options.

Sample Prompts

For AI-asks mode: "Let's play 20 Questions. I'm thinking of a historical figure from the American Revolution. Ask me yes-or-no questions to figure out who it is. After each of my answers, briefly explain your reasoning before asking the next question."

For student-asks mode: "Let's play 20 Questions. You are thinking of an element from the periodic table. I will ask yes-or-no questions to guess which element you've chosen. Answer only 'yes' or 'no.' Keep track of how many questions I've asked."

For clue mode: "I'm going to give you clues about something I'm thinking of. After each clue, try to guess what it is. Tell me how confident you are in your guess on a scale of 1 to 10."

Tracking Tools

- **Question log template:** A simple table with columns for question number, question text, answer, and remaining possibilities.
- **Information gain worksheet:** For each question, students estimate how many possibilities it eliminates.
- **Strategy reflection form:** Post-game questions for analysis. What was your best question? Your worst? What would you change?

The Real Work

Socrates believed that he knew nothing. This was not false modesty. It was a method. By claiming ignorance, he put himself in a position to ask questions. And through those questions, patiently pursued, he helped others discover what they knew but had not yet articulated.

The AI knows a great deal in one sense because it has processed vast quantities of text.It knows nothing for another reason, it has no understanding, no experience, no genuine curiosity about what you are thinking. It can simulate the question-asking process remarkably well, but it cannot actually wonder.

Your students can. They can wonder, and through wondering, they can learn to ask questions that cut through uncertainty and reveal what they want to know. The game of 20 Questions, played with an AI, becomes a training ground for this most human of skills.

Shannon showed that information is the reduction of uncertainty. Each good question reduces uncertainty by half. Students who internalize this principle, who learn to ask the questions that divide the world efficiently, are students who can think. The AI is the sparring partner. The thinking is theirs.

Example Prompts and Variations
Classic 20 Questions
Purpose: *Reveals AI reasoning patterns and question strategy*

I'm thinking of [CATEGORY: animal/historical figure/object/concept]. You have 20 yes-or-no questions to guess what it is. Ask one question at a time. After each answer, tell me your current best guesses and why.

Reverse 20 Questions
Purpose: *Develops student questioning skills and systematic thinking*

Think of a [CATEGORY]. I will ask you yes-or-no questions to guess what you're thinking of. Answer only yes or no. Keep track of how many questions it takes me.

The Taxonomy Game
Purpose: *Makes categorical thinking explicit and visible*

I'll give you clues about something, one at a time. After each clue, tell me: (1) What categories you've eliminated, (2) What categories remain possible, (3) What question would most efficiently narrow the remaining options. Clue 1: [FIRST CLUE]

Question Quality Analysis
Purpose: *Develops metacognition about questioning strategies*

I asked these questions to narrow down a mystery object: [LIST QUESTIONS]. Rate each question's efficiency on a scale of 1-10. Which questions split the remaining possibilities most evenly? Which were wasted?

Building a Decision Tree
Purpose: *Teaches systematic classification and binary search logic*

Help me create a decision tree to identify [CATEGORY OF THINGS]. What's the best first question to split all possibilities roughly in half? Then what's the best second question for each branch?

Chapter 5: Name That Plant

Phase	Time Required	Description & Activities
Class 1 (or Part A): Nature Walk	25–35 Minutes	Tech Check (3m): Apps ready. Field Work (15–25m): Identification using apps (e.g., Seek/iNaturalist). Return (5m): Organize findings.
Class 2 (or Part B): AI Investigation	15–20 Minutes	Deep Dive: Use AI to research the organism's life cycle, ecological role, and relationships within the local ecosystem.
Class 2/3: Report-Out	15–25 Minutes	Presentations: 1–2 minutes per student. Limit to 8–10 per day for large classes to maintain engagement.
Total Core Time	60–90 Minutes	Can be delivered as a single block or split into two 45-minute classes.
Extended: Phenology Project	45–60 Minutes	Seasonal Tracking: Repeat in Fall, Winter, and Spring to track changes over time.

The Use Case

Students use AI-powered identification apps to discover and learn about plants, trees, insects, and other organisms in their environment. After identifying something, they use AI chatbots to explore it more deeply, including its life cycle, role in the ecosystem, and relationships with other species.

The Pedagogy

In 2019, a research paper in Frontiers in Psychology made a provocative claim that time in nature may be essential for learning. The researchers reviewed evidence showing that access to natural spaces supports improved physical health and that time outside in nature has beneficial effects on cognitive and mental health. This was not news to teachers who had watched their students come alive on field trips, but it was the first time the research base had become substantial enough to take seriously.

A 2022 systematic review in Frontiers in Public Health analyzed 147 original research studies on nature-specific outdoor learning. The findings were striking: students who learned in natural outdoor settings showed increased engagement and ownership of their knowledge, evidence of academic improvement, development of social and collaborative skills, and improved self-concept. The researchers concluded that nature-specific outdoor learning should be incorporated into every child's school experience.

Most teachers are not trained naturalists. They cannot identify the plants in their schoolyard, much less explain the ecological relationships between species. How do you teach outdoor science when you don't know what you're looking at?

AI solves this problem. A teacher no longer needs to be an expert to lead students into nature. The technology has become remarkably accurate. Research published in 2019 found that Google Lens correctly identified plant species in 92.6% of cases, a result the researchers called "quite high" and sufficient to recommend the app for classroom use. The most excellent accuracy was observed when analyzing trees and plant stems. Even the identification of flowers, which is more variable, exceeded 90% accuracy under good conditions.

The pedagogical tradition here connects to what Louise Chawla calls "learning to love the natural world enough to protect it." Her research found that childhood experiences in nature, especially those involving discovery and exploration, are the strongest predictors of adult environmental concern. When a child identifies a plant for the first time, learns its name, and discovers its role in the web of life, something shifts. The anonymous green blur becomes a specific thing with a story.

AI makes this possible at scale. Every student can have their own moment of discovery. Every student can ask their own questions. The technology democratizes what was once the province of naturalists with years of training.

The Method
Phase 1: The Nature Walk and Identification
1. **Prepare the technology.** Ensure students have access to a smartphone or tablet with Google Lens, Seek by iNaturalist, or a similar identification app. Google Lens is pre-installed on most Android devices and available through the Google app on iOS. Seek by iNaturalist is free, privacy-focused, and explicitly designed for educational use.
2. **Define the boundaries.** Establish clear physical limits for the nature walk. The school grounds are often sufficient. A nearby park or natural area can expand possibilities. Students should stay within sight or designated zones.
3. **Give the assignment.** Each student (or pair) chooses one organism to investigate: a plant, tree, flower, or insect. They photograph it using the identification app, confirm the identification (or note if

uncertain), and record the common name and scientific name.
4. **Document the location.** Students note where they found their organism. This will become important in ecological discussions later. Was it in the sun or the shade? Near water? By the building or in the wild area?

Phase 2: The AI Investigation

Once students have identified their organism, they return to the classroom and use an AI chatbot to explore it more deeply. The prompts differ by grade level.

For High School Biology:
- What is the life cycle of this organism?
- What trophic level does it occupy?
- Where is it on the food chain? What eats it? What does it eat (or what nutrients does it require)?
- What are its natural predators?
- What is its ecological niche?
- How does this organism interact with other species in its ecosystem?

For Elementary Students:
- Tell me five cool things about this organism.
- Where does it live?
- What does it need to survive?
- Does it help other plants or animals? How?
- Is there anything surprising about it?

Phase 3: The Report-Out

Students report their findings to the class. Each presentation should include the organism's name (common and scientific), where it was found, one thing about its life cycle or needs, and one thing about how it relates to other organisms. The class builds a collective picture of the ecosystem right outside their door.

From the Field

The research on identification apps in classroom settings is growing. A 2022 study in Educational Technology Quarterly analyzed the didactics of using AR-recognition apps in biology classes and found that Google Lens is highly efficient for plant type and species identification. The researchers specifically noted that learning biology with Google Lens helps students construct their own knowledge, improving both digital literacy skills and learning outcomes.

The botany teacher at my school recently integrated AI technologies into our outdoor classroom activities, significantly enriching our plant identification lessons. Initially, students were introduced to basic prompts that guided them in recognizing various plant species, but as our lessons evolved, the complexity deepened. We began to explore specific topics such as trophic levels, illustrating how plants fit into broader ecological systems, and defense mechanisms, investigating how different species adapt to their environments to deter herbivores or pests.

To further enhance our learning experience, I successfully secured a grant to purchase AI glasses, which have proven to be an invaluable resource. These cutting-edge tools not only provided real-time data about the flora we encountered but also facilitated interactive learning in the field. Students were challenged to engage actively with the material; they couldn't simply copy and paste answers. Instead, they had to formulate relevant questions based on their observations and document their findings meticulously on clipboards as we explored the rich biodiversity of the woods and gardens surrounding our school. This hands-on approach fostered a deeper understanding of botanical concepts and encouraged critical thinking, making for a truly immersive educational experience.

The iNaturalist Educator's Guide recommends Seek by iNaturalist as a "fun, privacy-focused, and gamified app that provides live ID suggestions." The guide notes that Seek is particularly appropriate for younger students because it does not require posting observations to a public database. The California Academy of Sciences and the National Geographic Society developed Seek specifically for educational use, and it includes features such as badges and challenges that motivate students to find more species.

Typical Sense Education reviewed Seek and found that teachers can use it to "challenge students to photograph species of plants and animals that they regularly see around their neighborhood, and then identify them as a class." The review noted that having multiple students photograph the same organism from different angles often helps when one photo does not produce a match.

The National Wildlife Federation created a "Biodiversity at School" activity using Seek that includes a bingo card for students to fill with different categories of organisms. Students go on nature walks and try to complete their cards, combining the excitement of a scavenger hunt with genuine scientific observation.

The research consistently shows that these tools work best when combined with structured inquiry. Identification alone is not enough. Students need prompts that push them to think about relationships, cycles, and systems. This is where the combination of identification apps and AI chatbots becomes powerful; one provides the name, the other provides the context.

What Goes Wrong

- **The app cannot identify the organism.** Photo quality matters enormously. Blurry images, poor lighting, or photos taken from too far away will fail. Research shows that the app works best with close-up, well-lit images of distinctive features, such as flowers or leaves. Fix: Teach students to take multiple photos from different angles. If the app cannot make a match, that becomes a learning opportunity. Students can use the AI chatbot to describe what they see and ask for possible identifications based on characteristics and location.
- **The identification is wrong.** No app is 100% accurate. Google Lens achieves about 93% accuracy under good conditions, which means roughly 1 in 14 identifications may be incorrect. Fix: Treat identifications as hypotheses, not facts. Encourage students to cross-reference with field guides or a second app. Teach them to look for confirming evidence. This skepticism is itself a valuable scientific skill.
- **Students focus only on identification.** The identification becomes the goal rather than the doorway. Students collect names like Pokémon and never go deeper. Fix: Structure the assignment so that identification is only the first step. The AI investigation phase must be required, not optional. The report-out should emphasize relationships and systems, not just names.
- **Weather and timing problems.** Rain cancels the outdoor portion. The season is wrong for flowers. The schoolyard has been mowed flat. Fix: Plan alternatives. Students can bring in photos from home. The AI chatbot can still provide information about organisms in the local area even if students cannot observe them directly. A winter version might focus on tree bark identification or signs of animal activity.
- **Device and access inequity.** Not all students have smartphones. Some schools restrict device use. Fix: Pair students so that each pair has access to at least one device. Use school devices if available. Consider a station-based approach in which the nature walk occurs in groups that share devices.

Going Further

- **The Ecosystem Map.** Students plot their findings on a map of the school grounds, drawing arrows to indicate ecological

relationships, such as what eats what, what pollinates what, and what provides shelter for what. The map becomes a visual representation of the invisible web of connections.
- **The Phenology Project.** Repeat the activity across the school year. Track when different plants flower, when insects appear, and when leaves change. Students create a timeline of seasonal changes for their local ecosystem. This connects to climate science and helps students understand how timing matters in ecology.
- **The Invasive Species Hunt.** Focus specifically on identifying non-native or invasive species. Students use the AI to learn why particular species are problematic, what native species they displace, and what can be done. This connects to environmental policy and local conservation efforts.
- **The Food Web Challenge.** Working as a class, students attempt to construct a complete food web using only organisms they can identify on or near school grounds. How full can they make it? What is missing? What organisms might be present but hard to observe?
- **The Citizen Science Connection.** Students 13 and older can upload observations to iNaturalist, contributing to real scientific databases. Their observations become part of global biodiversity monitoring. Scientists actually use this data. Students are not just learning about science; they are doing science.

Your Toolkit
Identification Apps
- **Google Lens:** Pre-installed on most Android devices, available through the Google app on iOS. 92.6% accuracy in research studies. Identifies plants, animals, insects, and more. Free.
- **Seek by iNaturalist:** Privacy-focused, designed for education. Does not require posting to a public database. Includes badges and challenges. Created by the California Academy of Sciences and National Geographic. Free.
- **iNaturalist:** Full-featured citizen science app. Observations contribute to the global database. Community provides identification help. Requires account (13+). Free.
- **PlantNet:** Specialized for plant identification. Users specify the plant organ (leaf, flower, or fruit) to improve accuracy. Large database. Free.

Sample AI Prompts for Investigation
For ecology focus: "I found a [organism name] in [location]. What role

does it play in the local ecosystem? What other species depend on it, or does it depend on? What would happen if it disappeared?"

For life science focus: "Explain the complete life cycle of [organism name]. What does it need at each stage? How does it reproduce?"

For young learners: "Pretend you are a [organism name]. Tell me about your day. What do you eat? Where do you sleep? What are you afraid of?"

Additional Resources
- **iNaturalist Educator's Guide:** Comprehensive guidance for classroom use, including privacy considerations and pedagogical approaches.
- **National Geographic BioBlitz Guide:** Materials for organizing a BioBlitz (a focused biodiversity survey), aligned to Next Generation Science Standards.
- **National Wildlife Federation Seek Activities:** Ready-to-use classroom activities, including biodiversity bingo and nature walk challenges.

The Real Work

There is an old story about Louis Agassiz, the 19th-century naturalist. A student came to him wanting to study natural science. Agassiz handed him a fish and told him to observe it. The student looked at it for a while and said he was done. Agassiz told him to look again. This went on for three days. Only then did Agassiz explain that the student had finally begun to see.

The AI identification apps give students names. That is not the same as seeing. Names are powerful. They are the beginning of attention. Once you know that the tree by the parking lot is a red oak, you start to notice the shape of its leaves. You notice when the acorns drop. You see the squirrels gathering them. The name was the doorway.

The AI chatbot gives students context. That is still not the same as seeing. Context deepens attention. Once you know that red oaks can live 300 years, that their acorns feed dozens of species, that their roots connect with underground fungi in ways scientists are still learning, you look at that tree differently. It becomes a character in a story that extends far beyond what you can observe in an afternoon.

The beings who built Deep Thought never looked at the world around them. They wanted answers without observation. Your students, photographing plants and asking what they are, have already done more than those hyper-intelligent beings ever did. They have paid attention. This is what we want for our students, not just information, but the capacity to pay attention. The natural world is all around them, waiting to be noticed.

Example Prompts and Variations

Identification Verification

Purpose: Teaches verification of AI identifications against observable features

An AI identification app says this plant is [SPECIES NAME]. What key identifying features should I look for to confirm this? What similar species might be confused with it?

Habitat Context

Purpose: Connects individual observations to ecological understanding

I found a [SPECIES] in [LOCATION/HABITAT]. Is this typical for this species? What does finding it here tell me about the local ecosystem?

Field Guide Comparison

Purpose: Develops appreciation for different knowledge sources

Compare what an AI identification app might tell me about [SPECIES] versus what a traditional field guide would include. What information might be missing from each?

Observation Journal Entry

Purpose: Structures scientific observation and documentation

Help me write a naturalist's journal entry for this observation: Species: [NAME], Location: [PLACE], Date: [DATE], Conditions: [WEATHER/SEASON]. What details should I record? What questions should I investigate?

Ecosystem Connections

Purpose: Develop systems thinking about natural environments

I identified these species on my nature walk: [LIST]. What relationships might exist between them? What ecological story do these observations tell?

Chapter 6: Guess the Object

Activity Version	Time Required	Description & Activities
Class 1: Version 1	25–35 Minutes	Student Describes, AI Guesses: Select object (2m), write description (8–10m), AI testing (5m), analysis & revision (10–15m).
Class 1: Version 2	15–20 Minutes	AI Describes, Student Guesses: Receive clues one at a time and track how many are needed to identify the target.
Class 1: Version 3	30–40 Minutes	Five Senses Challenge: Multi-sensory brainstorming and writing. Best used with physical objects for tactile observation.
Class 1: Version 4	20–25 Minutes	Comparison Game: Focused on using AI to evaluate figurative language, metaphors, and simile construction.
Quick Version	5 Minutes	Daily Warm-up: Use Version 2 for subject-specific vocabulary at the start of every class.
Total Cycle Time	25–40 Minutes	A standard complete cycle of writing, testing, and revising.

The Use Case

Students describe objects to an AI and see if it can guess what they are describing. They can also reverse the game, the AI describes something, and students guess what it is. In both directions, the game demands precise language, careful observation, and attention to what makes one thing different from another.

This is a training ground for precision.

The Pedagogy

The game Taboo has been a staple of language classrooms for decades. In Taboo, players must describe a word without using certain "forbidden" words. If the target word is "elephant," you cannot say "trunk," "gray," "Africa," or "large." This constraint forces players to find alternative ways to express meaning, to reach for synonyms, to describe function and context rather than obvious attributes.

Research supports what teachers have observed. A 2023 study in the journal Ethical Lingua found that the Taboo game was perceived positively by student participants as a tool for vocabulary learning. The game assisted in "learning and understanding materials, making it easier to know and learn new vocabulary and increasing their learning creativity." A separate study found that teaching vocabulary with Taboo had "a positive effect on students' attitudes towards learning vocabulary, and on their interpersonal relationships." Both low-achieving and high-achieving students benefited equally.

The underlying principle is simple. When you cannot use the obvious word, you must think harder about what you actually mean. You must observe more carefully. You must find the precise language that captures the essential qualities of the thing you are describing.

This connects to a broader body of research on observation and descriptive writing. A study published in the International Journal of Science and Mathematics Education found that "before science can be completely understood, one of the fundamental skills that must be developed is observation." The researchers designed activities where students had to write descriptions so precise that another student could recreate what they had made. The gap between what students thought they had communicated and what was actually understood revealed how much more accurate their language needed to become.

Reading Rockets, a national literacy initiative, identifies the key elements of effective descriptive writing, which include vivid details that appeal to the senses, figurative language that creates connections, and precise word choice that eliminates ambiguity. They note that "descriptive writing is a skill, and a craft, that takes instruction, practice, and time to learn." The guessing game with AI provides precisely this kind of practice.

What AI adds to the traditional game is a patient, unbiased partner. The AI does not know what the student is thinking. It can only respond to the words on the screen. If the description is vague, the AI will guess wrong. If the description is precise, the AI will guess correctly. The feedback is immediate and inherent in the game itself. Did the AI understand you or not?

The Method
Version 1: Student Describes, AI Guesses
1. **Select the objects.** Prepare a set of objects or images for students to describe. These can be physical objects in the classroom, images on cards, or vocabulary words written on slips of paper. Choose items appropriate to your subject, such as science equipment, historical artifacts, literary symbols, mathematical shapes, or everyday objects.
2. **Set the rules.** Students cannot name the object directly. They cannot use the most obvious category word (for "apple," they cannot say "fruit"). They must describe using sensory details, function, comparisons, or context. You can add Taboo-style

restrictions by listing specific words they cannot use.
3. **Write the description.** Students write their description before engaging with the AI. This prevents them from adjusting on the fly. The description should be complete enough that someone who has never seen the object could identify it.
4. **Test with AI.** Students input their description to the AI with a prompt like: "I am going to describe something. Based on my description, what do you think it is? Here is my description [student's text]." The AI attempts to guess.
5. **Analyze and revise.** If the AI guessed correctly, the description worked. If not, students analyze why. What was missing? What was ambiguous? They revise and try again. The revision process is where the learning happens.

Version 2: AI Describes, Student Guesses
6. **Set up the prompt.** Tell the AI, "Think of a common household object. Describe it to me without naming it. Give me clues one at a time. Start with the most general clue and get more specific. Wait for me to guess after each clue."
7. **Play the guessing game.** Students receive clues one at a time and try to guess. They track how many clues they needed. The goal is to think of as few clues as possible.
8. **Analyze the clues.** After guessing (or giving up), students analyze the AI's descriptions. Which clues were most helpful? Which were too vague? What made a clue "good"? This builds vocabulary for discussing effective description.

Version 3: The Five Senses Challenge

Students must describe an object using all five senses: sight, sound, smell, taste, and touch. For objects that do not obviously engage certain senses (what does a chair sound like?), students must think creatively. The AI then guesses based on the sensory description. This version particularly develops the kind of vivid, concrete language that strengthens descriptive writing.

Version 4: The Comparison Game

Students describe objects using only comparisons, similes, metaphors, and analogies. "It is like a...," "It reminds me of...," "It functions similarly to...." This version builds figurative language skills and helps students see connections between disparate things.

From the Field

Edutopia, in its collection of language learning games, describes an activity called "Describe and Guess," in which one student selects an object from the classroom environment and describes it to the class without naming it. The goal is to use rich, specific details to help their classmates guess what it is." They recommend challenging students to use at least three adjectives in their description, building vocabulary while developing descriptive precision.

I have implemented a version of the Comparison Game numerous times in my classroom settings, and it has proven to be an invaluable tool across various science disciplines as well as in my manufacturing class. This interactive approach allows students to analyze and compare different materials for specific applications, fostering critical thinking and decision-making skills.

By facilitating side-by-side evaluations, the strengths and weaknesses of each material become evident in a short amount of time. This analytical process is further enhanced by integrating AI technologies that provide data-driven insights, enabling students to make informed choices grounded in real-world applications. As a result, we're able to minimize waste and significantly reduce costs associated with unused materials, creating a more sustainable and efficient learning environment. Overall, the Comparison Game not only enriches the educational experience but also prepares students for practical challenges in their future careers.

The research on vocabulary games extends beyond English language arts. A study on chemistry education found that Taboo-style games effectively helped students learn scientific terminology by forcing them to describe concepts in multiple ways. When students cannot simply recite a definition, they must truly understand what makes a term meaningful.

Hong Kong TESOL published guidance on using AI for vocabulary learning, including a guessing game format. They suggest prompting the AI, "I will show you a list of words. You need to choose one word on the list and give me clues to help me guess the word." The AI provides clues one at a time, and students must guess before receiving the next clue. This creates the same iterative, feedback-rich environment as the classroom version, but available for individual practice anytime.

Research on descriptive writing instruction consistently emphasizes the importance of feedback and revision. Reading Rockets notes that effective instruction includes "cycles of constructive teacher and peer feedback followed by thoughtful revision." The AI guessing game provides this feedback loop naturally; the AI's guess (correct or incorrect) is immediate feedback on the quality of the description. No teacher intervention is required to help students know whether their language was precise enough.

AI Lessons for the Classroom

What Goes Wrong
- **Descriptions are too short.** Students write "It is round and you eat it" and expect the AI to guess correctly. Fix: Set minimum requirements. Descriptions must include at least five distinct characteristics. Or require descriptions of a specific word count. The constraint forces elaboration.
- **Students use category words that give it away.** "It is a type of furniture" makes "chair" too easy. Fix: Add Taboo restrictions. Before starting, identify the two or three most obvious category words and ban them. This forces students to describe functions and attributes rather than classification.
- **The AI guesses correctly too easily.** Students describe everyday objects, and the AI gets them on the first try. Fix: Use more specific or unusual objects. Instead of "pencil," use "mechanical pencil" or "carpenter's pencil." Instead of "apple," use "Granny Smith apple" or "crabapple." The more specific the target, the more precise the description must be.
- **Students get frustrated when the AI guesses wrong.** They blame the AI rather than examine their description. Fix: Reframe failure as diagnostic information. When the AI guesses wrong, that is data about the description's effectiveness. Ask "What did the AI think you were describing? Why might it have made that mistake? What was missing from your description?"
- **The game becomes about tricking the AI.** Students write intentionally misleading descriptions. Fix: Clarify the goal. The purpose is not to stump the AI but to write a description so clear that the AI cannot fail to understand. Success is measured by how few words it takes to communicate precisely, not by how many wrong guesses you can induce.

Going Further
- **The Mystery Object Box.** Place an object in a box. Students reach in and feel it without looking. They must describe it using only touch. The AI guesses based solely on tactile description. This develops vocabulary for texture, temperature, weight, and shape.
- **The Abstract Concept Challenge.** Move beyond physical objects to abstract concepts such as justice, freedom, jealousy, and time. Students describe the concept without using the word or obvious synonyms. The AI attempts to identify the idea. This builds skills

for analytical and persuasive writing.
- **The Character Description.** For literature classes, students describe a character from a text they have read. The AI tries to identify the character. This test tests whether students can capture the essential qualities that distinguish one character from another.
- **The Process Description.** Students describe a process or procedure (how to tie a shoe, how photosynthesis works, how a bill becomes a law) and the AI identifies the process. This extends the game to expository writing and tests whether students can sequence steps clearly.
- **The Cross-Language Challenge.** For language learners, the game can be played in the target language. Students describe objects in Spanish, French, or Chinese. The AI guesses in the same language. This builds vocabulary and tests whether students can express meaning without resorting to their native language.

Your Toolkit
Sample Prompts for the AI

The student describes the mode: "I am going to describe something without naming it. Based on my description alone, what do you think it is? If you're not sure, make your best guess and explain your reasoning. Here is my description: [student text]."

For AI-describes mode: "Think of a [category: household object/animal/food / historical figure]. Give me clues one at a time, starting with the most general and getting more specific. After each clue, wait for me to guess before giving the next one."

For the five senses version: "I am going to describe something using my five senses. I will tell you what it looks like, sounds like, feels like, smells like, and tastes like. Based on these sensory details, tell me what you think it is."

For analysis mode: "Here is a description someone wrote of an object: [description]. The object was actually a [answer]. Analyze the description. What made it effective or ineffective? What details helped identify it? What was missing or confusing?"

Scaffolding Tools
- **Sense Chart:** A graphic organizer with five columns (sight, sound, smell, taste, touch) for students to brainstorm sensory details before writing.
- **Comparison Starter List:** Sentence frames like "It is shaped like...," "It reminds me of...," "It functions similarly to...," "It feels like..."

- **Taboo Word Cards:** For each object, a list of 3 to 5 words that cannot be used in the description. Students can also create these for each other.
- **Description Quality Rubric:** A simple rubric assessing: specificity (vague vs. precise), completeness (missing details vs. comprehensive), organization (scattered vs. logical), and language (common words vs. varied vocabulary).

Object Categories by Subject
- **Science:** Lab equipment, specimens, elements, compounds, biological structures, astronomical objects
- **History:** Artifacts, documents, historical figures, events, locations, inventions
- **Literature:** Characters, settings, symbols, themes, literary devices, authors
- **Math:** Shapes, operations, theorems, graphs, mathematical relationships
- **Art:** Artworks, art movements, techniques, materials, colors, compositions

The Real Work

There is a moment in every writer's development when they realize that the word they first reached for is not quite right. The apple is not just "red." It is crimson near the stem and fading to yellow at the bottom. It is not just "round." It is slightly lopsided, with a dimple where the stem attaches. The more carefully you look, the more you see. The more you see, the more precisely you can name what you see.

This is what the guessing game teaches. When the AI guesses wrong, it is because the student's words did not capture what made this thing this thing and not something else. The student must look again, think again, and find the word that draws the line between what something is and what it is not.

Mark Twain reportedly said that the difference between the almost-right word and the right word is the difference between the lightning bug and the lightning. The guessing game makes this difference visible. When "round" fails and "spherical" succeeds, when "big" fails and "roughly the size of a basketball" succeeds, students experience the power of precision.

The AI does not care about grades. It does not care about effort. It simply responds to the words it receives. In that simple response, whether it is a correct guess or an incorrect one, students find out whether their words were enough. The game is the teacher. The words are the lesson. The precision is theirs to develop.

Example Prompts and Variations

The Description Challenge
Purpose: *Develops precise descriptive language through immediate feedback*
I'm going to describe an object without naming it. After each sentence, tell me your best guesses, ranked by confidence. Here's my first description: [DESCRIPTION]

Sensory Description Practice
Purpose: *Expands descriptive vocabulary beyond visual observation*
I described something using only visual details. Now prompt me to add descriptions for other senses: touch, sound, smell, weight, and temperature. Ask me one sense at a time.

Specificity Coaching
Purpose: *Teaches the difference between general and specific descriptions*
My description: '[STUDENT DESCRIPTION]'. How could I make this more specific? What details are vague? Give me three ways to make the most generic phrase more precise.

Comparative Description
Purpose: *Develops analogical thinking and comparative language*
Help me describe [OBJECT] by comparison. What is it similar to? What is it different from? Complete this: 'It's like a ___ but ___.'

The Reverse Challenge
Purpose: *Models effective descriptive sequencing*
Describe an object for me to guess. Start with the most general feature and get progressively more specific. I'll tell you when I can imagine it. Let's see how few clues you need.

Technical vs. Poetic Description
Purpose: *Develops awareness of register and purpose in description*
Describe [OBJECT] two ways: first as a scientist would (precise, measurable, technical), then as a poet would (evocative, metaphorical, emotional). What does each approach capture that the other misses?

Chapter 7: Argue With Me

Activity Version	Time Required	Description & Activities
Class 1: Version 1	30–45 Minutes	AI as Practice Opponent: Position assignment (10m), AI sparring session (15–20m), and argument revision (10–15m).
Class 1: Version 2	30–45 Minutes	AI as Devil's Advocate: Integrates into existing discussion; adds a segment for AI perspective introduction and response.
Class 1: Version 3	20–30 Minutes	AI as Socratic Questioner: Individual AI engagement focusing on deepening student logic while peers observe or draft.
Class 1: Version 4	25–35 Minutes	Logical Fallacy Detection: Fallacy intro (10m), analyzing AI-generated fallacy examples (15–20m), and debrief (5–10m).
Full Debate Unit	3–4 Classes	Day 1: AI prep. Day 2: Steelmanning. Days 3–4: Live, non-AI debates.
Quick Version	10 Minutes	Anticipation Exercise Using AI-generated counterarguments before writing an essay.

The Use Case

Students prepare for debates by using AI as a practice opponent. The AI takes the opposing position, generates counterarguments, and challenges students to defend their views. In Socratic seminars, the AI can introduce perspectives that no student in the room holds, ensuring that all sides of an issue are represented.

This is about thinking, not winning.

The Pedagogy

The evidence for debate as a learning tool is robust. A 2024 study published in the American Educational Research Journal examined policy debate programs in Boston Public Schools serving economically disadvantaged students of color. The researchers found that debate had positive impacts on English Language Arts test scores of 0.13 standard deviations, equivalent to 68% of a full year of average ninth-grade learning. The gains were concentrated on analytical rather than rote subskills. The study also found evidence of positive effects on high school graduation and postsecondary enrollment. The most striking impacts were among students who were the lowest-achieving before joining the debate.

Harvard's ABLConnect database summarizes the research, "Debating is found to improve learning outcomes. The immediate positive effects include greater knowledge acquisition by reinforcing already taught materials." For controversial subjects in particular, "debating enhances students' appreciation for the complexities of the subject matter, and challenges prior beliefs." In the longer term, "debating helps students acquire better comprehension, application, and critical evaluation skills."

A 2021 intervention study published in System, an International Journal of Educational Technology and Applied Linguistics, examined debate instruction in Dutch secondary schools. The researchers found that debate instruction had positive effects on multiple structural components and quality aspects of both written and oral argumentation skills. They concluded that "debate constitutes a conducive pedagogy for honing argumentation skills."

The tradition here is ancient. Socrates, in 5th-century BCE Athens, developed a method of teaching through dialogue and questioning. Rather than delivering lectures, he engaged his students in conversation, probing their beliefs, exposing contradictions, and guiding them toward deeper understanding through argument. As Colorado State University's Institute for Learning and Teaching explains, "The Socratic Method is a dialogue between teacher and students, instigated by the continual probing questions of the teacher, in a concerted effort to explore the underlying beliefs that shape the students' views."

Modern Socratic seminars build on this tradition. ReadWriteThink, a project of the National Council of Teachers of English, describes the approach: "The Socratic seminar is a formal discussion, based on a text, in which the leader asks open-ended questions. Within the context of the discussion, students listen closely to the comments of others, thinking critically for themselves, and articulate their own thoughts and their responses to the thoughts of others."

Debate requires an opponent. Socratic seminars require diverse perspectives. In many classrooms, students may cluster around similar views, leaving important positions unrepresented. Or students may be unprepared to articulate the strongest version of an opposing argument.

AI can serve as that opponent. A 2023 study published in Computers & Education developed a novel task design called "chatbot-assisted in-class debates." Students interacted with an argumentative chatbot before engaging in debates with classmates. The chatbot helped students generate ideas for supporting their position and predict opposing viewpoints. The results showed that students had higher levels of enjoyment and exerted more effort when engaging in chatbot-assisted debates than when completing conventional learning tasks. The researchers found "empirical evidence that integrating argumentative chatbots into classroom debates can lead to improved argumentation skills and higher task motivation among undergraduate students."

Deep Thought never argued. It computed. It had no opponent to sharpen its thinking, no skeptic to expose its assumptions. Your students, debating with AI and with each other, are doing what the machine could not by testing ideas against resistance.

The Method
Version 1: AI as Practice Opponent
1. **Assign the position.** Give students a debatable proposition and assign them to argue one side. The proposition should be genuinely contestable, with reasonable arguments on both sides. Avoid topics where one side is obviously correct.
2. **Research and prepare.** Students research their position using traditional sources. They develop their main arguments, gather evidence, and anticipate counterarguments.
3. **Engage the AI opponent.** Students tell the AI, "I am going to argue that [position]. Your job is to argue the opposite. After I make an argument, respond with a counterargument. Challenge my reasoning. Point out weaknesses in my evidence. After each exchange, we will continue until I say stop."
4. **Practice the exchange.** Students present their arguments. The AI responds with counterarguments. Students must respond in real time, just as they would in an actual debate. This builds the skill of thinking on your feet.
5. **Analyze and revise.** After the practice session, students review the exchange. What arguments did they struggle to rebut? What evidence did the AI cite that they had not considered? They revise their preparation accordingly.

Version 2: AI as Devil's Advocate in Discussion
In a class discussion where all students seem to agree, the teacher can introduce the AI as a dissenting voice. Project the AI's responses on screen so the whole class can see. The AI argues a position that no student is defending, forcing students to engage with perspectives they might otherwise dismiss. This ensures that even in a room full of like-minded students, the strongest opposing arguments get heard.

Version 3: AI as Socratic Questioner
Rather than taking a position, the AI can be prompted to ask probing questions. "I will state my position on an issue. Your job is not to agree or disagree but to ask me questions that challenge my thinking, expose

assumptions I might be making, and push me to consider implications I might not have thought about." This replicates the Socratic method; the AI leads through questions rather than assertions.

Version 4: Logical Fallacy Detection

Students can use AI to learn to identify rhetorical tricks and logical fallacies. Research published in PLOS ONE found that teaching students to recognize informal fallacies can help inoculate them against misinformation. The study concluded that "interventions that teach how to identify informal fallacies correctly could play a role in mitigating the effects of misinformation." Common fallacies students should learn to recognize include

- **Ad hominem:** Attacking the person making the argument rather than the argument itself.
- **Straw man:** Misrepresenting an opponent's argument to make it easier to attack.
- **Appeal to emotion:** Manipulating feelings instead of presenting logical reasoning.
- **False dichotomy:** Presenting only two options when more exist.
- **Slippery slope:** Claiming that one action will inevitably lead to extreme consequences.
- **Appeal to authority:** Citing an authority figure whose expertise is irrelevant to the topic.

Students can prompt the AI to intentionally use these fallacies, then practice identifying them. Or they can submit arguments and ask the AI to identify any fallacies present.

From the Field

Times Higher Education published guidance on using AI in classroom debates, based on an Executive MBA module at a major university. The instructor used ChatGPT in the "constructive phase" of debate preparation, where students build arguments supported by data, case studies, and theoretical frameworks. The AI assisted with "fact-checking, data collection, and argument formulation." Crucially, AI was excluded from the "rebuttal phase," where students must respond spontaneously to opponents' arguments. As the instructor explained, "Real-time decision-making, understanding nuanced arguments and employing persuasive tactics cannot yet (and should not) be fully replicated by AI."

One of my personal favorites for engaging 9th graders is the Devil's Advocate concept. This approach resonates well with them, as they have a natural inclination to debate and scrutinize the nuances of language and meaning. However, they often find themselves sidetracked from the main argument, which can derail the discussion. By using AI to serve as Devil's Advocate, the students are forced to stay focused and on topic. The AI's ability to present counterarguments in a structured manner encourages it to refine its thoughts and articulate its points more clearly, ultimately enhancing the quality of the discourse. This method not only sharpens their critical thinking skills but also promotes a deeper understanding of the subject matter.

The results were notable "The use of ChatGPT in the constructive phase of debate planning has led to a marked improvement in the quality of structured debates; students employed a broader array of case studies and presented a more diverse range of arguments." However, the instructor also observed that "some students took certain information at face value without adequately scrutinising its merit." These unexamined assumptions were then "dissected during the cross-examination phase and, most notably, in the impromptu rebuttals, for which no preparation time and AI are allowed."

Research from the University of Mississippi, published in Argumentation and Advocacy, argues that debate builds skills that are particularly valuable in the age of AI. The researchers note that "recent studies show that using AI tools to assist with writing results in a 25% reduction in accuracy and a 12% decline in reading comprehension." Debate, they argue, provides an antidote. "We saw that debate offers skills that could make students better able to navigate those problems." As one researcher explained, "Debate's not just about stating the points. You've got to be able to respond in a time-pressured setting to what your opponent is going to say and rebuild your arguments, and then compare and weigh arguments. Those are things that a college debater will, over time, become very good at, and I do not think the current generation of AI platforms is doing quite as well."

Edutopia published guidance on Socratic seminars that emphasizes the importance of student ownership "This type of student-led discussion, based on Socrates' method of student inquiry rather than teacher lecture, elicits student ownership, deep thinking, critical questioning, academic vocabulary usage, and a rooted sense of community." The key is that the teacher "steps aside" once students are prepared. AI can help with that preparation, ensuring students have considered all perspectives before the discussion begins.

What Goes Wrong
- **Students rely on AI arguments instead of developing their own.**

The AI becomes a script rather than a sparring partner. Fix: Require students to develop their initial arguments before engaging with the AI. The AI session should test and refine existing arguments, not generate them from scratch.
- **The AI is too easy to beat.** Students point out weak arguments and feel overconfident. Fix: Prompt the AI to be a strong opponent. "Make the strongest possible case for the opposing position. Do not concede points easily. Challenge every claim I make." You can also prompt it to cite specific evidence and studies.
- **The AI makes up facts.** AI can generate plausible-sounding but false citations. Fix: Teach students to verify any claims the AI makes. This is itself a valuable skill. The AI's potential to hallucinate becomes a lesson in source verification.
- **Students become defensive instead of curious.** The practice session turns into an ego battle rather than a genuine inquiry. Fix: Reframe the goal. The purpose is not to "win" against the AI but to discover weaknesses in your own argument. Every strong counterargument the AI raises is a gift, showing you where you need more preparation.
- **The AI refuses to argue a position.** Some AI systems are reluctant to argue for positions that seem harmful or unethical. Fix: Frame requests carefully. "For educational purposes, help me understand the strongest arguments someone might make for this position." Or switch to a different topic where the AI is willing to engage.

Going Further
- **The Debate Tournament.** Hold a classroom debate where both sides have used AI to prepare. In the actual discussion, no AI assistance is allowed. Students must rely on what they learned and internalized during preparation. Award points for arguments that successfully rebut points the AI had raised.
- **The Argument Map.** After debating with the AI, students create a visual map of the argument that includes the central claims, supporting evidence, counterarguments, and responses to counterarguments. This makes the structure of reasoning visible and helps students see where their argument is strongest and weakest.
- **The Steelman Challenge.** Students must write the strongest possible version of the opposing argument, the "steelman" rather than the "straw man." They use the AI to help develop this, then

present it to the class. Points are awarded for how convincingly they argue against their own position. This builds intellectual humility and genuine understanding of opposing views.
- **The Fallacy Hunt.** Collect examples of arguments from news, social media, or advertising. Students analyze these with AI assistance to identify logical fallacies. Create a classroom gallery of fallacies in the wild to build media literacy and critical thinking simultaneously.
- **The Historical Debate.** Stage a debate between historical figures or perspectives. The AI can role-play as a historical figure, arguing from their documented positions and using language appropriate to their era. Students must engage with historical arguments on their own terms rather than dismissing them with the benefit of modern hindsight.

Your Toolkit
Sample AI Prompts
For debate practice: "I am preparing for a debate. My position is [X]. Your job is to argue the opposite as strongly as possible. After I present each argument, respond with your strongest counterargument. Challenge my evidence, question my reasoning, and force me to defend every claim. Do not concede points easily."

For Socratic questioning: "I will share my view on a topic. Rather than agreeing or disagreeing, ask me probing questions that challenge my assumptions, expose potential contradictions, and push me to think more deeply. Do not give me answers. Only ask questions."

For fallacy detection: "I will share an argument. Analyze it for logical fallacies. If you find any, name them, explain why they are fallacious, and suggest how the argument could be strengthened by removing the fallacy."

For steelman construction: "I hold [position X]. Help me understand the strongest possible case for the opposing view. What are the best arguments someone might make against my position? What evidence supports that view? What would a reasonable, intelligent person who disagrees with me say?"

For argument analysis: "Here is an argument I am making: [argument]. Identify the weakest points. What claims need more evidence? What counterarguments am I not addressing? Where could an opponent attack this argument most effectively?"

Debate Topics by Subject
- **Science:** Should genetic engineering of humans be permitted? Is nuclear power the best solution to climate change? Should we

colonize Mars?
- **History:** Was the American Revolution justified? Did the ends justify the means in [historical event]? How should we judge historical figures by modern standards?
- **Literature:** Is [character] a hero or a villain? Does the author's intent matter in interpreting a text? Should controversial books be taught in schools?
- **Ethics:** Classic trolley problems, duties to future generations, individual rights versus collective good
- **Current Events:** Policy debates on technology, environment, education, economics (choose topics with genuine disagreement among reasonable people)

Resources for Debate Education
- **ReadWriteThink Socratic Seminars Guide:** Practical methods for implementing Socratic seminars in the classroom.
- **Stanford National Forensic Institute:** Research and resources on the benefits of debate education.
- **Kialo Edu:** Free platform for structured classroom discussions with argument mapping.
- **National Paideia Center:** Resources for implementing Socratic seminars based on Mortimer Adler's approach.

The Real Work

In Plato's dialogues, Socrates often claims that he knows nothing. This is a method. By acknowledging uncertainty, Socrates opens space for genuine inquiry. He does not lecture. He questions. He does not tell students what to think. He helps them discover the limits and contradictions of what they already think.

The AI is not Socrates. It has no genuine curiosity, no fundamental uncertainty, no stake in the outcome. It can serve a Socratic function; it can generate the questions, the counterarguments, the alternative perspectives that students need to encounter. It can be the voice that says, "But have you considered..." when no one else in the room is willing to.

The danger is mistaking the practice for the thing itself. Debating with an AI is a practice. The honest debate happens between humans, in real time, with real stakes. The real thinking happens inside the student's mind when they encounter an objection they cannot immediately answer. The real growth occurs when someone changes their mind, not because they lost an argument but because they learned something.

Example Prompts and Variations

Steelmanning the Opposition
Purpose: *Develops intellectual honesty and prepares for real opposition*
I'm arguing that [POSITION]. Present the strongest possible case against my position. Don't use weak arguments. Give me the version that would be hardest for me to refute.

Anticipating Rebuttals
Purpose: *Builds debate preparation and defensive argumentation*
My argument: [ARGUMENT]. What are the three most likely rebuttals I'll face? For each, give me a potential response I could use.

Logical Fallacy Detection
Purpose: *Teaches recognition of flawed reasoning*
Analyze this argument for logical fallacies: [ARGUMENT]. Name each fallacy you find, explain why it's a fallacy, and show how the argument could be made stronger without the fallacy.

The Socratic Opponent
Purpose: *Models Socratic questioning for student practice*
I'm going to make a claim. Your job is to ask me questions that reveal weaknesses or unexamined assumptions in my reasoning. Don't argue with me. Just ask probing questions. My claim: [CLAIM]

Evidence Evaluation
Purpose: *Develop a critical evaluation of supporting material*
I'm using this evidence to support my argument: [EVIDENCE]. How strong is this evidence? What would make it stronger? What counter-evidence should I be aware of?

Perspective Shift
Purpose: *Builds empathy and multi-perspective thinking*
I'm debating [TOPIC] from the perspective of [STAKEHOLDER A]. Now, argue the same topic from the standpoint of [STAKEHOLDER B]. What values and priorities shift? Where do they actually agree?

Chapter 8: Group Chat

Phase	Time Required	Description & Activities
Class 1: Version 1	60–90 Minutes	Complete Cycle: Dilemma intro/Persona assignment (10m), AI persona research (20m), Group prep (15m), Council discussion (30m), Debrief (15m).
Class 1: Version 2	+5 Minutes	AI as Dilemma Generator: AI generates a custom ethical scenario at the start before standard timing begins.
Class 1: Version 3	+10 Minutes	AI as Synthesizer: Add time at the end for AI to summarize the debate and for students to critique what it missed.
Class 1: Version 4	+5–10 Minutes	AI as Absent Stakeholder: Use AI to provide a "missing" perspective mid-discussion (e.g., a person from the future or a local animal).
Two-Day Option	2 x 45 Minutes	Day 1: Intro, Persona Research, & Prep. Day 2: Council Discussion, Cross-examination, & Debrief.
Compressed	45 Minutes	Reduce the number of personas and the length of the cross-examination period.

The Use Case

Students work in groups while using AI as a catalyst for discussion. The AI poses dilemmas, role-plays from different perspectives, or summarizes ideas for teams. Each group can receive a different AI persona to represent in a "Council of Thinkers" scenario, where students must understand and advocate for viewpoints they did not initially hold.

This is about multiplying perspectives.

The Pedagogy

Collaborative learning has decades of research behind it. A study published in CBE Life Sciences Education examined factors that make group work effective in higher education. The researchers found that successful collaboration required "student autonomy and self-regulatory behavior, combined with a challenging, open, and complex group task that required the students to create something new and original." When these conditions were met, students developed "a sense of responsibility and of shared ownership of both the collaborative process and the end product."

However, the researchers also identified a standard failure mode in which students divided the work rather than truly collaborating. As the study notes, Johnson and Johnson refer to groups showing this behavior as "pseudo learning groups." The result is "a lack of synthesis" that disappoints both teachers and students. The extent to which students benefit from working with others "depends on the quality of their interactions."

This is where AI can help. A 2025 study published in the International Journal of Educational Technology in Higher Education investigated the effects of generative AI on collaborative problem-solving and team creativity. Students who used AI tools like ChatGPT during group work reported that the tools "facilitated more group behaviors by providing immediate feedback and content generation." One student explained, "Midjourney gave us more opportunities to discuss in groups." Another noted, "With GAI tools, our group discussions became more constructive, and everyone was more engaged."

The study found that AI fostered "ongoing reflective process, enabling students to evaluate and adjust their decisions regularly." As one student said, "The instant feedback from ChatGPT made us rethink our choices. It wasn't just about accepting ChatGPT's suggestion but discussing it and deciding whether it made sense for our project."

Role-playing is another powerful pedagogical tool that AI can enhance. Research published by Harvard's Instructional Moves program explains that "having students take a position they don't necessarily agree with can make discussions less personal and invite broader participation." When students are freed from defending their own views, they can explore ideas more openly.

A study published in Science and Engineering Ethics found that role-play in education "not only aims at teaching students to become aware of the different dimensions in decision making, it also encourages students to think about what such an institutional framework for responsible action might look like." The researchers noted that it "often helps to play such role plays twice. In the first instance, the participants will keep up appearances. In the second instance, the group can be asked explicitly from now on to enter some serious negotiations to reach an agreement."

The Journal of Educational Research published a study on the effect of role-play on 21st-century skills. The researchers found that "in role-play, students are often required to embody characters and perspectives different from their own, encouraging them to explore relationships, meanings, and materials." This, at its best, "facilitates both content delivery as well as students' natural playful and social behavior."

What happens when we combine collaborative learning, role-playing, and AI? We get the "Council of Thinkers," a structured discussion where AI provides each group with a distinct perspective to understand, internalize, and represent. The AI is not replacing the student's voice. It is expanding the range of voices available in the room.

The Method
Version 1: The Council of Thinkers
1. **Choose the dilemma.** Select a complex question with multiple legitimate perspectives. This could be an ethical dilemma, a policy decision, a historical turning point, or a scientific controversy. The question should have no easy answer and should genuinely benefit from multiple viewpoints.
2. **Assign the personas.** Divide the class into groups of three to five students. Assign each group a different "thinker" to represent. These could be historical figures, stakeholder groups, philosophical schools, or fictional perspectives. Each persona should have a distinct viewpoint on the dilemma.
3. **Research with AI assistance.** Each group uses the AI to understand their assigned perspective. They might prompt "Explain how [persona] would view [dilemma]. What values would they prioritize? What arguments would they make? What would they say to someone who disagreed?" The AI provides the intellectual foundation; students build understanding on top of it.
4. **Prepare the position.** Groups work together to internalize their persona's perspective. They should be able to explain the viewpoint, defend it against objections, and identify its strengths and limitations. The goal is genuine understanding, not caricature.
5. **Convene the council.** Bring the groups together for a structured discussion. Each group presents their persona's perspective on the dilemma. Other groups can ask questions and raise objections. The goal is not to "win" but to illuminate the full complexity of the issue.
6. **Debrief and synthesize.** After the council, step out of character. Discuss what was learned. Where did perspectives overlap? Where were they genuinely incompatible? What did students learn about the issue that they did not know before?

Version 2: AI as Dilemma Generator
Before the group discussion begins, use the AI to generate a dilemma tailored to your content. Prompt: "Generate an ethical dilemma related to [topic] that would challenge high school students. The dilemma should have at least three possible responses, each with legitimate arguments for

and against. Include enough context for students to discuss meaningfully." The AI creates the scenario; students work in groups to analyze and respond.

Version 3: AI as Discussion Synthesizer

After a group discussion, have students summarize their conversation by inputting key points to the AI. Prompt: "We just discussed [topic]. Here are the main points that were raised: [points]. Please synthesize these into a coherent summary, identify any contradictions or tensions, and suggest questions we should explore further." The AI helps students see patterns in their own thinking and identify gaps worth pursuing.

Version 4: AI as Absent Stakeholder

In any discussion, some perspectives are likely to be underrepresented. The AI can speak for those absent voices. If discussing environmental policy, the AI might represent future generations. If debating a historical decision, the AI might represent the voices of those whose voices were not recorded in historical documents. The AI ensures that even perspectives no one in the room holds get a hearing.

Sample Personas for the Council

For Ethical Dilemmas:

- **The Utilitarian:** Seeks the greatest good for the greatest number. Focuses on outcomes and consequences.
- **The Deontologist:** Focuses on rules and duties. Some actions are right or wrong regardless of consequences.
- **The Virtue Ethicist:** Asks what a person of good character would do. Focuses on cultivating virtues.
- **The Care Ethicist:** Prioritizes relationships and responsibilities to specific others over abstract principles.
- **The Pragmatist:** Focuses on what works in practice. Skeptical of rigid theoretical frameworks.

For Historical Decisions:

- **The Contemporary Critic:** Someone who opposed the decision at the time and can articulate why.
- **The Decision Maker:** The person who made the decision, explaining their reasoning and constraints.
- **The Affected Party:** Someone directly impacted by the decision, for better or worse.
- **The Modern Historian:** Someone with hindsight, able to see long-term consequences.
- **The Foreign Observer:** Someone from another culture or nation, with a different frame of reference.

For Policy Debates:
- **The Economist:** Focuses on costs, incentives, and unintended consequences.
- **The Ethicist:** Focuses on rights, justice, and moral obligations.
- **The Practitioner:** Someone who would implement the policy, concerned with feasibility.
- **The Affected Community:** Those who would live with the policy's effects daily.
- **The Future Generation:** Someone considering long-term effects on those not yet born.

From the Field

Benjamin Breen, a history professor, has written extensively about using ChatGPT as a "history simulator" in the classroom. He developed what he calls "History Lens," an assignment in which students interact with AI that simulates historical settings. Student feedback was enthusiastic: "The plague simulation History Lens assignment was a great project that allowed us to experience what life was like during the time." Another student noted, "The instructor helped me feel engaged with the course very frequently because he used assignments and activities to allow the class to not just learn about history but to let us see through the eyes of the people during that time."

MIT's Scheller Teacher Education Program is researching "Collaborative Artificial Intelligence for Learning." The project focuses on "integrating innovative AI tools in the classroom to support learning in collaborative groups." Researchers explain that "students interact with AI-powered conversational agents in group work and discussions. The agent is envisioned as a peer who would promote deeper thinking on the topics instead of an efficiency tool."

EdTechTeacher published guidance on using ChatGPT for role-playing activities, noting that "role-playing activities can engage students in active learning and require them to think critically." They suggest having ChatGPT "adopt the role of different characters or historical figures relevant to the subject taught. For example, a history lesson can take on a famous leader's persona, allowing students to interact with and ask questions to gain deeper insights into their perspectives, decisions, and historical events."

In my engineering and manufacturing course, we have consistently utilized artificial intelligence as a hypothetical stakeholder during each semester when constructing our stakeholder matrices. By integrating AI into these matrices, we explore its potential impacts, benefits, and challenges in engineering processes. This approach allows us to thoughtfully assess the multifaceted roles that AI can play within our projects. We view these matrices as essential frameworks for driving substantial, long-term improvements in systems design and operational efficiency. Through this method, we not only emphasize the importance of considering all relevant stakeholders but also highlight the transformative role that AI can have in enhancing our engineering practices.

Cornell University's Center for Teaching Innovation provides extensive guidance on collaborative learning. They note that "collaborative work allows students to serve as thought partners for their peers to make sense of what they are learning, clarify misconceptions, and deepen their understanding." Students also "develop a more nuanced and complex understanding from exposure to multiple perspectives."

AVID Open Access published guidance on AI and collaboration, addressing the apparent tension between AI and human teamwork. They suggest using AI to generate ideas that groups then critically evaluate. "Once the AI has returned a list of ideas based on the prompt, students can work together to review the suggestions and determine which ideas are actually worth pursuing. This is a critical step in the process of using AI and a great collaborative thinking activity as well."

What Goes Wrong

- **Groups defer to the AI instead of discussing among themselves.** Students ask the AI every question rather than working through ideas together. Fix: Structure the activity with clear phases. AI consultation happens during research. Group discussion happens separately, without AI. The council convenes with AI turned off. Build in protected time for human-only conversation.
- **Students parrot the AI's words without internalizing the perspective.** They can recite the position but cannot explain it in their own words or respond to challenges. Fix: Require students to demonstrate their assigned perspective without notes before the council convenes. If they cannot explain it, they do not yet understand it.
- **The personas become caricatures.** A "utilitarian" is reduced to "doesn't care about individuals." A "conservative" is reduced to stereotypes. Fix: Emphasize that steelmanning students must present the strongest, most sympathetic version of their assigned perspective. If other groups can easily dismiss their arguments, they have not done the work.

- **Dominant students take over the group.** Some students do all the talking while others remain silent. Fix: Assign specific roles within groups: researcher, spokesperson, questioner, and note-taker. Rotate roles between activities. Require input from each group member during the council.
- **The discussion never reaches synthesis.** Groups present their positions but never engage with each other. Fix: Structure the council to require engagement. After the presentations, have a round in which each group asks another group a genuine question. Then have groups identify where they might find common ground. End by asking, "What would a solution that all these perspectives could accept look like?"

Going Further
- **The Rotating Perspective.** Rather than each group representing one perspective throughout, have groups rotate perspectives mid-discussion. After presenting as the Utilitarian, a group must now argue as the Care Ethicist. This prevents identification with a single viewpoint and builds genuine flexibility of thought.
- **The Synthesis Challenge.** After the council, the groups reform with one member from each original group. These new mixed groups must create a synthesis position that incorporates insights from all perspectives. They present how each perspective contributed to the synthesis.
- **The Silent Council.** Conduct the council in writing rather than speaking. Groups exchange written position papers, then written responses. This gives quieter students more time to think and can produce more careful, nuanced engagement with others' ideas.
- **The Real-World Extension.** After the council, students research how the dilemma was actually resolved (if historical) or how it is currently being debated (if contemporary). They compare their council's discussion to the real-world discourse. What perspectives were missing? What insights did students have that real-world participants missed?
- **The Student-Created Council.** Have students design their own council activity for younger students or for another class. They must identify a dilemma, create personas, write AI prompts to help groups understand each perspective, and design discussion protocols. Teaching the method deepens understanding of it.

Your Toolkit
Sample AI Prompts

For persona research: "I need to understand the perspective of [persona] on [dilemma]. Explain their worldview, their core values, and how they would approach this issue. What arguments would they find most compelling? What would they say to someone who disagreed with them? Help me understand this perspective deeply enough to represent it fairly."

For historical figures: "Act as [historical figure]. I am going to ask you questions about [topic]. Respond as that person would, based on their documented views, writings, and the context of their time. If I ask something they never addressed, explain how their known views might apply."

For dilemma generation: "Create an ethical dilemma suitable for high school students studying [topic]. The dilemma should have at least four possible responses, each with legitimate arguments. Make it complex enough that reasonable people could disagree but concrete enough that students can discuss specific options."

For discussion synthesis: "Our group just discussed [topic]. Here are the main points raised: [list]. Please identify the key areas of agreement, the main points of disagreement, any assumptions that went unchallenged, and questions we should explore further."

For absent stakeholders: "We are discussing [policy/decision]. Represent the perspective of [absent group, e.g., future generations, those without political voice, affected communities]. What would they want us to consider? What might they say if they could be part of this discussion?"

Discussion Protocols
- **Timed rounds:** Each group has three minutes to present their perspective. A two-minute questioning period follows. A timer keeps things moving.
- **Speaking tokens:** Each student has a limited number of tokens to "spend" on speaking. This prevents domination by vocal students and encourages quieter students to participate.
- **The "yes, and" rule:** Before critiquing another perspective, you must first acknowledge what is valuable about it. This prevents discussions from becoming purely adversarial.
- **The "what would change your mind" question:** Each group must articulate what evidence or argument would change their assigned perspective's mind. This tests genuine understanding.
- **The silence before synthesis:** Before attempting to synthesize perspectives, have two minutes of silence for individual reflection. This prevents the first idea offered from dominating.

The Real Work

The novelist F. Scott Fitzgerald is often credited with saying that "the test of a first-rate intelligence is the ability to hold two opposed ideas in mind at the same time and still retain the ability to function." The Council of Thinkers tests this ability directly. Students must understand a perspective well enough to represent it, even if they do not personally hold it. They must engage with opposing views seriously enough to respond to them, even if they do not find them convincing.

The goal is not to conclude that all perspectives are equally valid. Some arguments are stronger than others. Some positions rest on better evidence. Some solutions work better in practice. You cannot evaluate perspectives you do not understand. And you cannot understand perspectives you have never genuinely tried to inhabit.

The AI provides the raw material for articulating perspectives that students might not encounter in their daily lives. The AI cannot understand. It cannot hold two ideas in mind simultaneously. It cannot feel the tension between competing goods. It cannot make a judgment about what matters most.

That work remains human. The AI multiplies the voices in the room. The students must decide which voices to heed.

Example Prompts and Variations

The Council of Thinkers
Purpose: *Creates AI personas for multi-perspective discussions*
You are [HISTORICAL/FICTIONAL FIGURE]. Stay in character throughout our discussion. We're going to discuss [TOPIC]. Respond as this person would, drawing on their known beliefs, experiences, and ways of thinking.

Discussion Starter
Purpose: *Creates substantive discussion material*
Generate a dilemma about [TOPIC] that has no straightforward correct answer. The dilemma should force people to weigh competing values. Include enough detail that a group could discuss it for 15 minutes.

Idea Synthesis
Purpose: *Helps groups find common ground and clarify differences*
Our group discussed [TOPIC]. Here are the different positions that emerged: [LIST POSITIONS]. What do these positions have in common? Where are the genuine disagreements? What might a synthesis look like?

Devil's Advocate
Purpose: *Prevents groupthink and encourages critical examination*
Our group is leaning toward [CONCLUSION]. Before we finalize this, present the strongest case for why we might be wrong. What are we not seeing? What could go wrong?

Perspective Generator
Purpose: *Ensures diverse viewpoints are considered*
We're discussing [ISSUE]. Generate five different stakeholder perspectives we should consider. For each, explain: Who are they? What do they value? What would they want from this decision?

Discussion Debrief
Purpose: *Identifies gaps and extends learning beyond the discussion*
Our group discussion covered these points: [SUMMARY]. What important aspects of [TOPIC] did we not address? What questions remain unresolved? What should we explore next?

Chapter 9: Build a Model

Activity Version	Time Required	Description & Activities
Quick Challenge	45–50 Minutes	Intro (5m): Challenge launch. AI Consult (8m): Strategy/design search. Cycle 1 (15m): Build & test. PDSA Reflection (5m): Analyze results. Cycle 2 (15m): Final build & debrief.
Standard Format	90 Minutes (2 Classes)	Class 1: Intro, AI design consultation, and Build Attempt 1. Class 2: Review data, run PDSA Cycles 2 & 3, and perform final class-wide comparison.
Extended Format	135–225 Minutes (3–5 Classes)	Rube Goldberg Style: Each class period focuses on designing, testing, and perfecting a single stage of a complex chain reaction.
PDSA Cycle Speed	12–18 Minutes per Cycle	Includes Plan (AI/Human design), Do (Build), Study (Test), and Act (Refine). Aim for 2–3 cycles minimum.

The Use Case

Students design, build, and test physical solutions to constrained challenges. They might build a bridge from limited materials, design a Rube Goldberg machine, or create an eco-village layout. AI serves as a planning partner, a source of ideas, a feasibility checker, or a judge. The building happens with hands. The testing occurs in the real world. And the iteration happens through trial, failure, and improvement.

This is where AI meets its limits. And that is precisely the point.

The Pedagogy

Design thinking has been embraced across educational contexts for its emphasis on iteration, prototyping, and learning from failure. A study published in the Journal of Engineering Education examined the impact of teaching iterative prototyping to middle school students. The researchers found that "design thinking, with its emphasis on iterative prototyping and mantra of 'fail early and often,' stands in stark contrast to the typical one-and-done, failure-averse culture of the classroom."

The study's results were compelling. Students who received instruction on the iterative prototyping process and mindset "reported more positive affect and actions in reaction to failure and produced more successful designs." The researchers concluded that "instruction on the iterative prototyping process and mindset can encourage students to try early and often and promote healthier reactions to failure." Students who tested their designs earlier in the process created more successful final products.

The Plan-Do-Study-Act cycle, developed by W. Edwards Deming, provides a structured framework for this iterative process. The Deming Institute describes it as "a systematic process for gaining valuable learning and knowledge for the continual improvement of a product, process, or service." The cycle has four steps: Plan (formulate a theory and predict outcomes), Do (test the change), Study (compare outcomes to predictions), and Act (integrate learning and adjust).

In education, PDSA cycles have proven powerful. An article in Edutopia describes how teachers use PDSA cycles to "create positive change" in their classrooms. The approach gives "students more ownership over their learning" by involving them in identifying problems, proposing solutions, testing changes, and evaluating results. The Carnegie Foundation has championed PDSA as a core tool of improvement science in education, arguing that it provides the "disciplined inquiry" needed to learn what works in specific contexts.

Project-based learning research supports this approach. A study published in Computer Applications in Engineering Education found that integrating design thinking with project-based learning led to "greater motivation and creativity in comparison with those who followed traditional teaching methods." The International Journal of STEM Education published research showing that "applying the engineering design process to STEM project-based learning is beneficial for developing preservice technology teachers' schema of design thinking, especially with respect to clarifying the problem, generating ideas, and testing and revising solutions."

Wikipedia's entry on design-based learning summarizes the research "Positive benefits of the design-based learning approach have been observed, including student-based learning where students (often) identify their project's needs, develop their own ideas, and engage in a larger range of thinking than with the traditional scripted inquiry model." A 2000 study found that "the design project led to better learning outcomes and included deeper learning than the traditional learning approach."

AI cannot do what students must do in these challenges. Research on AI and creativity is revealing essential limitations. A study published in Scientific Reports compared human and AI creativity on divergent thinking tasks. The researchers found that "on average, the AI chatbots outperformed human participants." However, "the best human ideas still matched or exceeded those of the chatbots." AI produces consistently good ideas; humans make the occasional breakthrough.

Research from Harvard Business School found a similar pattern. When comparing AI and human solutions to business problems, evaluators judged the human solutions as "more novel, employing more unique 'out of the box' thinking." The AI-generated ideas were "more valuable and feasible" but less original. A study from Wharton found that while AI helps individuals produce better ideas, it reduces diversity across a group. In this study, only 6% of the AI-generated ideas were considered unique, compared with 100% of the ideas generated by humans.

Most significantly, a 2025 study published in Frontiers in Psychology found that "AI struggles to apply knowledge flexibly" and "frequently generated unrealistic or incorrect solutions" when faced with real-world problem-solving tasks that require both divergent and convergent thinking. The researchers concluded that "while AI can mimic human creativity, its strong performance in creative tasks is likely driven by non-creative mechanisms rather than genuine creative thinking."

This is why hands-on challenges matter. AI can suggest solutions. It cannot build them, test them, feel them fail, revise them based on what the failure revealed, and try again. That learning happens only through doing.

From the Field

I am a devoted fan of the BBC show Taskmaster. Comedians compete in absurd challenges with strange constraints. Build the tallest tower using only your body. Get a coconut as far from the starting point as possible. Make the best sandwich while wearing oven mitts. The joy is in watching people think creatively within limits, fail spectacularly, and occasionally discover brilliant solutions no one anticipated.

I brought this format into my Introduction to Engineering classroom. Every week, students competed in Taskmaster-style challenges. Toss a teabag the farthest. Build the tallest tower using only paper. Support the largest stack of books with a single piece of paper held one inch off the table or ground.

The last challenge became particularly interesting when we introduced AI. I asked ChatGPT to propose solutions for supporting the maximum weight with a single piece of paper elevated one inch above the surface. The AI produced reasonable suggestions such as rolling the paper into a tight cylinder, folding it into an accordion pattern, and creating a triangular prism. All were sensible and based on known principles of structural engineering.

Then my students ran the PDSA cycle. They started with the AI's suggestions and tested them. The cylinder held some books, but it buckled.

The accordion compressed. The triangular prism performed reasonably but reached a limit. Each failure generated learning. Each test revealed something the AI's theoretical knowledge had missed about how paper actually behaves under load, how the table surface affects stability, and how the books distribute weight unevenly.

After several cycles of Plan-Do-Study-Act, students began developing hybrid solutions that the AI had not proposed. They combined elements from different approaches. They discovered that the specific paper we were using had particular properties that changed the optimal design. They found that the one-inch height constraint created unexpected leverage challenges. By the third or fourth iteration, student designs were outperforming everything the AI had suggested.

The AI could not beat what the students came up with. Not because students were smarter than the AI in any abstract sense, but because students could do what the AI could not. They built, tested, felt the failure, understood what the failure meant, and revised based on actual physical experience. The AI possessed knowledge, while the students developed know-how.

That distinction matters. Knowledge is information about how things work. Know-how is the ability to make things work in specific contexts with specific materials under particular constraints. Knowledge can be transmitted. Know-how must be developed through practice. The AI transmitted knowledge. The PDSA cycle developed know-how.

The Method
Version 1: AI as Planning Partner
1. **Present the challenge.** Give students a constrained design problem with clear success criteria. The best challenges have simple goals, limited materials, and objective measurements. Examples: tallest free-standing structure, longest bridge span, fastest marble run, heaviest load supported.
2. **Consult the AI.** Before building, students ask the AI for suggestions. Prompt: "I need to [goal] using only [materials] with these constraints: [constraints]. What design approaches might work? Explain the physics or engineering principles behind each suggestion."
3. **Build and test.** Students build their prototype based on AI suggestions and their own ideas. Test against the success criteria and record the results.
4. **Run the PDSA cycle.** This is where learning happens. Plan: Based on what failed, what will you change? Predict what will happen.

AI Lessons for the Classroom   83

Do: Build the revised prototype. Study: Test it. Did it perform as predicted? What surprised you? Act: What did you learn? What will you try next?

5. **Document the iteration.** Students keep an engineering notebook recording each cycle: what they predicted, what happened, what they learned, and what they changed. This documentation is as essential as the final product.

Version 2: AI as Feasibility Judge

After students develop their design plans, they submit them to the AI for feasibility analysis. Prompt: "I'm planning to build [design] using [materials] to achieve [goal]. Analyze this design. What are the potential failure points? What physics principles am I relying on? What might go wrong that I haven't anticipated?" Students must then address the AI's concerns or explain why they disagree. The AI serves as a critical friend, forcing students to defend their thinking before they build.

Version 3: AI as Improvement Consultant

After a failed test, students describe the failure to the AI and ask for diagnostic help. Prompt: "My [design] failed when [description of failure]. Here's what I observed: [observations]. What might have caused this? What modifications might address this problem?" The AI provides hypotheses; students test them. This teaches that AI is a tool for generating possibilities, not a source of definitive answers.

Version 4: Rube Goldberg Challenge

Rube Goldberg machines are chain-reaction contraptions that accomplish simple tasks through elaborate sequences. A study published in Contemporary Issues in Technology and Teacher Education found that Rube Goldberg machines "have a positive influence on the STEM awareness of prospective science teachers." TeachEngineering provides a detailed curriculum noting that these projects teach "the engineering design process" and require students to "define the problem, gather information, brainstorm ideas, select the most promising idea, build and test, and redesign for improvement."

For the AI-assisted version: Students design their machine sequence with AI consultation at each stage. They ask the AI about physics principles, potential failure points, and alternative approaches. They build with physical materials, test in the real world, and iterate based on actual results. The chain reaction either works or it does not. No amount of AI consultation changes that physical reality.

The PDSA Framework

Each iteration should follow a structured cycle:
1. **Plan.** What change are you making? Why? What do you predict will happen? Write down your prediction before testing. This forces explicit thinking and creates something to compare against.
2. **Do.** Implement the change. Build the prototype. Run the test. Document what actually happens, not what you expected to happen.
3. **Study.** Compare results to predictions. What matched? What surprised you? Why might the unexpected results have occurred? This is where learning happens. Failure that matches the prediction teaches something. Failure that surprises teaches more.
4. **Act.** Based on what you learned, what will you do differently next time? This might mean adopting the change, adapting it, or abandoning it entirely. Each decision should be based on evidence from the Study phase.

The cycle repeats until time runs out or the goal is achieved. The value is not just in the final product but in the documented learning across iterations.

What Goes Wrong

- **Students skip the planning phase.** They want to build immediately without thinking through predictions. Fix: Require written predictions before any building. Make the prediction explicit and public. Students who skip this step miss the most valuable learning opportunity, comparing expectations with reality.
- **Students give up after the first failure.** The failure-averse classroom culture makes students want to abandon approaches that do not work immediately. Fix: Normalize failure explicitly. Share stories of famous engineering failures. Require multiple iterations as part of the assignment. Grade the quality of iteration, not just the final result.
- **AI becomes a crutch rather than a tool.** Students follow AI suggestions without critical evaluation or personal insight. Fix: Require students to explain why they are implementing any AI suggestion. Ask "What do you think will happen and why?" If they cannot explain the physics or logic, they should not be building it.
- **The struggle gets removed.** Teachers or AI provide so much guidance that students never experience productive frustration. Fix: Resist the urge to help too quickly. Let students sit with

problems. The struggle is the learning. If you remove the struggle, you remove the growth.
- **Documentation becomes busywork.** Engineering notebooks are filled in after the fact without genuine reflection. Fix: Build documentation into the workflow. Pause between each cycle for students to write. Make documentation part of the grade that matters. Review notebooks during the project, not just at the end.

Going Further
- **The Constraint Escalation.** After students succeed at a challenge, add a new constraint. Your tower must also hold weight. Your bridge must now span a longer distance. Your marble run must include a loop. Each constraint forces a new iteration and reveals new learning.
- **The Material Swap.** Students who mastered a challenge with one material must now achieve it with different materials. Paper becomes straws. Tape becomes string. This teaches that solutions are material-specific and that knowledge must adapt to context.
- **The Teaching Challenge.** Successful students must teach their approach to others without building it themselves. This tests whether they truly understand the principles or just stumbled upon a solution. Teaching reveals gaps in understanding.
- **The AI Competition.** Pit AI against students explicitly. Give the AI the same challenge and the same constraints. Compare the AI's best suggestion to what students produce after three PDSA cycles. Document where AI succeeded, where students succeeded, and what each approach revealed about the problem.
- **The Failure Museum.** Create a display of the most instructive failures from the project. What did each failure teach? Students present their failures to the class and explain what they learned. This celebrates failure as a source of learning rather than something to hide.

Your Toolkit
Sample Challenges
- **Tower Challenge:** Build the tallest free-standing structure using 20 sheets of paper and 1 meter of tape. Must stand for 10 seconds.
- **Bridge Challenge:** Span a 30cm gap using only popsicle sticks and glue. Measure the maximum weight supported before failure.
- **Egg Drop Challenge:** Protect an egg from a drop of 3 meters using materials that weigh less than 50 grams.
- **Marble Run Challenge:** Create the longest travel time for a

marble using cardboard and tape. Marble must be in motion the entire time.
- **Paper Support Challenge:** Support the maximum weight using a single piece of paper held 1 inch off the surface.

Sample AI Prompts

For initial design: "I need to build [goal] using only [materials] with these constraints: [constraints]. What design approaches might work? For each suggestion, explain the physics or engineering principles that make it work."

For feasibility check: "Here is my design plan: [description]. Analyze this design. What are the potential failure points? What am I assuming that might not be true? What could go wrong?"

For failure analysis: "My [design] failed when [description]. I observed [observations]. What might have caused this failure? What modifications might address this problem?"

For iteration guidance: "I have tried [approaches] and achieved [results]. Each failed because [reasons]. What other approaches might I consider that address these failure modes?"

Engineering Notebook Template

Each PDSA cycle should include: Date and iteration number. Design sketch with labels. Prediction (what will happen and why). Test results (what actually happened). Comparison (how did reality differ from prediction?). Learning (what did this teach me?). Next steps (what will I change and why?).

Additional resources: TeachEngineering (teachengineering.org), Rube Goldberg Institute (rubegoldberg.org), PBLWorks (pblworks.org), Arizona STEM Acceleration Project (stemteachers.asu.edu).

The Real Work

There is a difference between knowing and doing. AI excels at knowing. It can access vast stores of information about physics, engineering, materials science, and structural design. It can explain principles clearly. It can suggest approaches based on patterns it has learned from countless examples.

It cannot build. It cannot feel the paper buckle under its fingers. It cannot sense the moment when a structure begins to sway. It cannot experience the frustration of a design that worked in theory but failed in practice, then channel that frustration into the curiosity that leads to a better understanding. I've seen this gap in AI in my engineering and manufacturing classes. Even the virtual reality welding machines only get my students 70% of the way there. Students are surprised by the heat, sparks, dimming of the welding mask, and the pressure of the wire against the steel when they do the actual work.

These hands-on challenges reveal AI's limitations not as a weakness of technology but as a reminder of what human learning requires. We do not learn to ride a bicycle by reading about balance. We do not know how to cook by memorizing recipes. We do not develop engineering judgment by consulting an AI about structural principles.

We learn by doing it. By failing. By studying what went wrong. By trying again with a new understanding. The PDSA cycle is not just a pedagogical technique. It describes how human beings develop expertise in any domain.

AI can accelerate this process by providing better starting points and more informed hypotheses. It can serve as a knowledgeable consultant during the Study phase, helping students understand why things failed. It cannot build. It cannot do the testing. It cannot develop the tacit knowledge that comes only from hands engaged with materials in the world.

When students build a paper structure that outperforms the AI's suggestions, they have learned something no amount of information can teach: that knowledge is not enough. That context matters because the gap between theory and practice is where expertise develops. And that they are capable of producing that expertise themselves.

The beings in Douglas Adams's novel made a mistake that engineers learn to avoid. They built the most powerful computer in the universe and asked it an abstract question What is the meaning of life, the universe, and everything? They did not prototype. They did not test. They did not iterate. They waited 7.5 million years for an answer they could not use because they had never clarified what they were actually asking.

Deep Thought, to its credit, diagnosed the problem. "I think the problem, to be quite honest with you," it said, "is that you've never actually known what the question is." The machine recommended building another computer, the Earth, to determine which question would make the answer meaningful. The program would run for ten million years. Its processors would be the living beings who inhabited it, stumbling through existence, making mistakes, learning from failure.

Your students, building bridges from paper and towers from straws, are doing what the beings in Adams's novel refused to do. They are testing their ideas against reality. They are discovering, through failure, what questions they should have been asking. They are learning that the answer is never the point. The purpose of this process is to determine the answer you truly need.

The AI can tell them what a cantilever bridge is. It cannot tell them why their cantilever bridge collapsed. Only the collapse itself teaches that. Only the rebuilding cements the lesson. In Adams's universe, the Earth was a computer designed to process questions through lived experience. Your classroom serves the same function for a few hours each week.

Example Prompts and Variations

Initial Design Consultation
Purpose: *Generates starting points while teaching underlying principles*
I need to build [STRUCTURE/DEVICE] using only [MATERIALS]. The constraints are: [CONSTRAINTS]. What are three different design approaches I could try? For each, explain the key principle that makes it work.

Feasibility Analysis
Purpose: *Develop predictive thinking before building*
Here's my design plan: [DESCRIPTION]. Analyze its feasibility. What are the likely failure points? What physics or engineering principles am I relying on? What should I test first?

Failure Diagnosis
Purpose: *Transforms failure into a learning opportunity*
My design failed. Here's what happened: [DESCRIPTION OF FAILURE]. Why do you think it failed? What does this failure tell me about the underlying principles? What should I try differently?

Iteration Guidance
Purpose: *Supports PDSA cycle thinking*
In my last test, [WHAT HAPPENED]. My prediction was [WHAT I EXPECTED]. The difference tells me [MY ANALYSIS]. What modification would you suggest I try next, and why?

Rube Goldberg Planning
Purpose: *Supports complex chain-reaction design*
I'm building a Rube Goldberg machine. Stage [N] needs to [ACTION] and trigger stage [N+1], which needs to [NEXT ACTION]. What mechanisms could connect these stages? What's most likely to fail?

Constraint Introduction
Purpose: *Develops adaptability and principle-based thinking*
My design works with [CURRENT MATERIALS]. Now I need to achieve the same goal using [NEW CONSTRAINT]. What principles from my original design still apply? What needs to change completely?

Design Documentation
Purpose: *Develops engineering notebook and reflection skills*
Help me document my design process. I started with [INITIAL IDEA], tested [WHAT I TESTED], learned [WHAT I LEARNED], and ended with [FINAL DESIGN]. What's the story of how iteration improved my result?

Chapter 10: Quick Wins

Activity	Time	Best For	Research Basis
What Did I Just Say?	5-8 min	Introducing new concepts	Retrieval practice
Three-Word Summary	5-7 min	Reading comprehension	Summarization
Better Question	8-12 min	Building AI literacy	Prompt engineering
Confidence Calibration	10-15 min	Exam preparation	Metacognition
Socratic Interrupt	8-10 min	Critical analysis	Socratic questioning
Error Hunt	10-12 min	Verification skills	Active learning
Translation Challenge	8-12 min	Deep comprehension	Teaching to learn

It is Sunday night. You have picked up this book looking for something practical. The activities in the main chapters are powerful, but they require planning: printing rubrics, preparing AI prompts, setting up verification stations. You do not have that kind of time right now.

This appendix is for you.

The seven activities that follow require no preparation beyond having access to an AI chatbot. Each takes between five and fifteen minutes. Each can be adapted to any subject. And each builds genuine learning skills rather than mere AI fluency.

These are not shortcuts. They are distillations. Behind each activity lies substantial research on how learning actually works. The brevity is the point.

The Research Foundation

Quick does not mean shallow. These activities draw on three decades of cognitive science research, distilled into forms that work in the time you actually have.

Retrieval practice is the finding that actively recalling information strengthens memory more than passive review. Roediger and Karpicke (2006) demonstrated that students who practiced retrieving information retained significantly more than those who simply reread material, even when the rereading students felt more confident in their learning. The testing effect, as this phenomenon is called, has been replicated across hundreds of studies. Roediger, Agarwal, McDaniel, and McDermott (2011) found that quizzing in middle school classrooms improved exam performance by the equivalent of a full letter grade.

Summarization forces students to identify essential information and discard the peripheral. Haystead and Marzano (2009) found that summarizing strategies increased students' understanding of content by an average of 19 percentile points across 17 experimental studies. The constraint of reducing text to its essence requires precisely the kind of active processing that transfers information from working memory to long-term storage.

Metacognition refers to thinking about one's own thinking. The Education Endowment Foundation rates metacognition and self-regulation as having high impact for relatively low cost, with research showing that students who develop metacognitive skills can accelerate their learning by the equivalent of seven additional months of progress. Schraw and Moshman (1995) established that metacognitive knowledge, including awareness of one's own learning processes, predicts academic success across domains.

Socratic questioning stimulates critical thinking through systematic inquiry. Research from Colorado State University's Institute for Learning and Teaching confirms that "the Socratic Method is a dialogue between teacher and students, instigated by the continual probing questions of the teacher, in a concerted effort to explore the underlying beliefs that shape the students' views." This approach has been shown to develop analytical skills more effectively than direct instruction alone (Oyler & Romanelli, 2014).

Prompt engineering is an emerging skill that research suggests has direct educational benefits. Sperling and colleagues (2024) found that students who received structured prompt training significantly outperformed those without such training on data analysis and programming tasks across all levels of Bloom's taxonomy. Critically, higher-quality prompting skills predicted higher-quality AI outputs, suggesting that prompt engineering is indeed a teachable, valuable skill.

Each activity below applies one or more of these principles. The research is robust. The execution is simple.

Activity 1: The "What Did I Just Say?" Check
Time: 5-8 minutes | Materials: AI chatbot, any topic | Research basis: Retrieval practice, self-explanation

The Setup
Have AI explain a concept from your current unit. Students read or listen to the explanation. Then they close the AI window and explain the concept back in their own words, either in writing or to a partner.

Why It Works
This activity exploits the testing effect: the act of retrieving information strengthens memory more than passive review. When students read an AI explanation and then try to reproduce it, they quickly discover what they actually understood versus what merely seemed familiar. Research by Roediger and Karpicke (2006) found that retrieval practice produced "substantially greater retention than studying" even when students who merely studied felt more confident about their learning.

The Prompt
> *Explain [CONCEPT] to a high school student in about 150 words. Use clear language and one concrete example.*

Variations
- Partner exchange: Students explain to each other without notes, then compare their versions
- Accuracy check: After students write their explanations, they reopen the AI and compare, noting what they missed
- Teaching test: Students explain to an imaginary younger sibling who knows nothing about the topic

Activity 2: The Three-Word Summary

Time: 5-7 minutes | Materials: AI chatbot, any text or topic | Research basis: Summarization, metacognition

The Setup
AI generates a paragraph of explanation on your current topic. Students must reduce that paragraph to exactly three essential words. Not a three-word sentence. Three isolated words that capture the core meaning.

Why It Works
The extreme constraint forces genuine prioritization. Students cannot include everything, so they must decide what matters most. Marzano and colleagues found that summarization strategies significantly improve comprehension, but only when students must "discern the inherent structures in a text" rather than simply deciding what seems important. The three-word limit makes this discrimination unavoidable. Students who struggle to choose their three words are revealing gaps in understanding.

The Prompt
> *Write a 100-word paragraph explaining [TOPIC]. Make it informative but do not include bold text or bullet points.*

Variations
- Defend your choices: Students must justify why each word made the cut
- Compare and contrast: Different students will choose different words; discussing why reveals different understandings
- Progressive reduction: Start with ten words, then five, then three

Activity 3: The "Better Question" Challenge
Time: 8-12 minutes | Materials: AI chatbot | Research basis: Prompt engineering, metacognition

The Setup
Display an AI response to a vague question. Students must craft a better, more specific prompt that would produce a more useful response. They test their improved prompts and compare results.

Why It Works
Prompt engineering research shows that the quality of AI output directly correlates with prompt quality. Sperling et al. (2024) found that "higher-quality prompt engineering skills predict the quality of LLM output," making this a genuinely teachable skill with immediate, visible feedback. More importantly, crafting precise prompts requires the same cognitive work as clear thinking: identifying what you actually want to know, what context matters, and what constraints would be helpful. Lo (2023) describes this as "prompt literacy," a new foundational skill for the information age.

Example Sequence
Vague prompt: "Tell me about photosynthesis."
Improved prompt: "Explain the light-dependent reactions of photosynthesis to a 10th-grade biology student. Focus on what happens in the thylakoid membrane and explain why water molecules are split."

Variations
- Prompt competition: Students vote on whose improved prompt produced the most useful response
- Constraint challenge: Add specific requirements (word count, reading level, format) to prompts
- Audience shift: Rewrite prompts for different audiences (expert, child, skeptic)

Activity 4: The Confidence Calibration
Time: 10-15 minutes | Materials: AI chatbot, index cards or paper | Research basis: Metacognition, retrieval practice

The Setup
AI generates five factual statements about your current topic, a mix of accurate and inaccurate claims. Students rate their confidence in each

statement (1-5 scale) before checking. After verification, they compare their confidence ratings to actual accuracy.

Why It Works

Metacognitive research consistently shows that students are poor judges of their own knowledge. Dunlosky and Rawson (2012) found that "overconfidence produces underachievement" because students who believe they understand material do not invest effort in actual learning. This activity makes the gap between perceived and actual knowledge visible. The Education Endowment Foundation notes that "improved metacognition and self-regulation skills have the potential to promote learning across the curriculum" with an effect size equivalent to seven months of additional progress.

The Prompt

> Generate five factual statements about [TOPIC]. Make three of them accurate and two of them subtly incorrect. Do not indicate which are which.

Variations

- Pattern tracking: Over multiple sessions, students track their calibration accuracy and identify blind spots
- Source hunt: Students must find sources that confirm or refute each statement
- Discussion pairs: Students compare confidence ratings before checking, discussing why they disagree

Activity 5: The Socratic Interrupt

Time: 8-10 minutes | Materials: AI chatbot | Research basis: Socratic questioning, critical thinking

The Setup

AI provides an explanation. At two or three key points, you pause the reading and ask students: "What assumption is the AI making here?" or "What question should we ask before accepting this?" Students generate questions before continuing.

Why It Works

Socratic questioning develops critical thinking by making implicit assumptions explicit. Research from the University of Pittsburgh confirms that "pursuing clarity of expression through careful follow-up questions" and "searching for the evidence and reasons" are key components of productive questioning. The Wikipedia entry on Socratic questioning notes that "the level of thinking that occurs is influenced by the level of questions asked." By interrupting AI explanations for questioning, students practice the skeptical stance that serves them in all information environments.

AI Lessons for the Classroom

The Prompt
> *Explain [CONCEPT] step by step. After each major point, pause and write [PAUSE] so I can ask questions before you continue.*

Sample Interruption Questions
- What evidence supports this claim?
- What alternative explanation might there be?
- What is the AI assuming we already know?
- Who might disagree with this, and why?

Activity 6: The Error Hunt

Time: 10-12 minutes | Materials: AI chatbot | Research basis: Active learning, verification skills

The Setup
Tell the AI to intentionally include errors in its explanation. Students must find them. The activity transforms passive reading into active investigation.

Why It Works
Error detection requires deeper processing than simple comprehension. Students must compare each claim against their existing knowledge and notice inconsistencies. Research on AI hallucinations shows that errors often appear in plausible-sounding forms, which is precisely why students need practice identifying them. This activity also normalizes the idea that AI output requires verification, building the healthy skepticism that Chapter 10 describes as essential.

The Prompt
> *Write a 150-word explanation of [TOPIC] that is mostly accurate but contains exactly two factual errors. Do not identify where the errors are.*

Variations
- Unknown count: Do not tell students how many errors exist
- Error types: Request specific error categories (wrong dates, incorrect cause-and-effect, misattributed quotes)
- Difficulty scaling: Start with obvious errors, progress to subtle ones

Activity 7: The Translation Challenge

Time: 8-12 minutes | Materials: AI chatbot | Research basis: Comprehension, audience awareness

The Setup
AI explains a concept at an advanced level. Students must "translate" it for a specific audience: a younger sibling, a grandparent, someone from a different field. The translation cannot simply be shorter; it must make the concept genuinely accessible while remaining accurate.

Why It Works

Translation for different audiences requires deep comprehension. You cannot simplify what you do not understand. Research on the "protege effect" shows that preparing to teach material produces better learning than preparing to be tested on it. When students translate complex AI explanations for different audiences, they engage in exactly this kind of teaching-oriented processing. The MIT Teaching and Learning Lab notes that "experts possess more knowledge that is better organized and integrated," and translation exercises build this organizational capacity.

The Prompt

> *Explain [CONCEPT] at a college level, using technical vocabulary and assuming significant background knowledge.*

Translation Targets

- A curious eight-year-old
- A grandparent who never studied this subject
- A friend who thinks this topic is boring
- A newspaper headline writer (25 words maximum)

A Note on Adaptation

These activities are frameworks, not scripts. A chemistry teacher might use the Three-Word Summary with AI explanations of chemical bonding. A history teacher might use the Error Hunt with AI-generated historical narratives. An English teacher might use the Translation Challenge with AI analysis of literary themes.

The principle remains constant: AI provides the raw material. Students do the cognitive work. Learning happens in the transformation, not the transmission.

It is still Sunday night. You now have options for tomorrow morning. Use one. See what happens. Iterate from there.

Chapter 11: When the Machine Breaks

On September 26, 1983, Lieutenant Colonel Stanislav Petrov sat at the command center of the Soviet early-warning satellite system. The bunker, known as Serpukhov-15, monitored American nuclear activity. Shortly after midnight, the system's alarms triggered. Five intercontinental ballistic missiles had been launched from the United States toward the Soviet Union. The screen flashed "LAUNCH" with a "high reliability" indicator.

The protocol required immediate notification up the chain of command. A retaliatory strike would follow within minutes. The system was confident. The system was unambiguous. The system was telling Petrov that nuclear war had begun.

Petrov did not follow protocol.

He reasoned that if the United States were launching a first strike, it would involve hundreds of missiles simultaneously to disable Soviet counterattack capability, not just five. The satellite system was new. It had experienced glitches before. Ground radar showed nothing. Something did not add up.

He reported the alert as a false alarm and waited.

The missiles never came. Investigation later determined that a rare alignment of sunlight on high-altitude clouds, combined with the satellites' Molniya orbits, had triggered the false positive. The system had been confident. The system had been wrong. Petrov's willingness to distrust that confidence may have prevented nuclear war.

The incident remained classified for over a decade. When it finally became public in the 1990s, Petrov was asked how he had known the system was wrong. His answer was simple; he had a funny feeling in his gut. That feeling was the accumulated judgment of a career spent understanding both the capabilities and the limitations of the technology he operated. He trusted his training more than he trusted the machine.

This chapter is about AI failure. It is really a chapter about judgment, the human capacity to recognize when a confident system is producing nonsense, and to know what to do about it.

The Pedagogy

Every technology fails. What matters is whether users understand the failure modes well enough to catch problems before they cascade.

Research on what psychologists call "automation bias" demonstrates this challenge precisely. In a landmark 1998 study, Kathleen Mosier and colleagues at NASA Ames Research Center examined how pilots interact with automated cockpit systems. They found that pilots using automated decision aids committed both "omission errors" (failing to notice problems because computerized systems did not flag them) and "commission errors" (following automated recommendations even when other cockpit instruments provided contradictory information).

The findings were striking. In one experiment involving a simulated engine fire with conflicting cues about which engine was affected, 75% of pilots followed the electronic checklist's wrong recommendation to shut down the unaffected engine. Pilots using traditional paper checklists committed this error only 25% of the time. The automation was helpful often enough that vigilance eroded.

Mosier and Skitka, writing in the *International Journal of Aviation Psychology*, defined automation bias as "errors made when decision makers rely on automated cues as a heuristic replacement for vigilant information seeking and processing." The term captures something essential: when a system is usually right, we stop checking whether it is right this time.

Your students face the same challenge. AI is right often enough that trust builds. Then it fabricates a source, miscounts syllables, or contradicts itself mid-paragraph. Students who have not learned to watch for these failures pass them along as truth.

The solution is not to avoid AI. It is to study its failure modes systematically, the way a pilot studies what can go wrong with an aircraft. Not to create fear, but to develop competence. The student who has seen AI hallucinate a citation will check citations forever after. The student who has caught AI in a confident mathematical error will verify calculations as a matter of habit. These are not unfortunate side effects of AI use. They are among the most valuable lessons AI can teach.

The Method

This chapter presents a taxonomy of AI failures. Each failure type is a potential lesson, a diagnostic opportunity, a moment when technology's limitations illuminate something important about thinking itself.

For each failure type, you will find:
- **What It Looks Like:** How to recognize this failure when it appears
- **The Research:** What studies have documented about this failure mode
- **Why It Happens:** The technical or design reason behind the breakdown
- **In the Classroom:** How to turn this failure into learning

Failure Type One: The Invented Source
What It Looks Like
A student asks AI for sources on a research topic. The AI provides what appears to be a properly formatted citation, author name, article title, journal, volume, page numbers, and year. The citation looks precisely like a real one. It is not real. The article does not exist.
The Research
A 2023 study published in *Scientific Reports* by William H. Walters and Esther Isabelle Wilder systematically tested citation accuracy in AI-generated content. The researchers examined 636 bibliographic citations generated by ChatGPT across 84 documents on multidisciplinary topics. The findings were stark 55% of citations produced by GPT-3.5 were entirely fabricated. Even the more advanced GPT-4 fabricated 18% of its citations. Among the real citations that GPT-3.5 produced, 43% contained substantive errors. GPT-4's error rate on real citations was 24%.

These fabricated citations were not random nonsense. They included real author names, properly formatted digital object identifiers, and plausible journal titles. They looked exactly like real citations. They just happened to reference papers that were never written.

Jordan MacDonald, a doctoral student at the University of New Brunswick, conducted a focused study on ChatGPT's citations in psychology research and published it in *Mind Pad*, a journal of the Canadian Psychological Association. He found that 32.3% of 300 citations were hallucinated. The rate varied dramatically by subfield, with only 6% in neuropsychology and 60% in psychology of religion. "Almost every single citation had hallucinated elements or was just entirely fake," MacDonald told PsyPost, "but ChatGPT would offer summaries of this fake research that were convincing and well worded."

A 2024 study at the University of Mississippi, examining student-submitted work, found that 47% of citations from AI sources had incorrect titles, dates, authors, or combinations of all three. The students were not being dishonest. They did not know the AI was making things up.
Why It Happens
AI does not retrieve citations from a database. It generates text based on patterns learned from training data. It has learned what citations look like, the format, the typical structure, and the kinds of words that appear in academic article titles. When asked for sources, it produces text that matches those patterns.

The AI is not malfunctioning. It is doing precisely what it always does, generating statistically probable text. The problem is that statistically probable text and factually accurate text are not the same thing. A citation that looks right is not the same as a proper citation.

This is why ChatGPT doubled down when Schwartz asked whether the fake cases were real. The model was not lying in any intentional sense. It was generating what a confident, reassuring response would look like. Confident responses say "Yes, this exists." Reassuring responses provide details about where to find it. The model provided both, because that is what fluent text on this topic looks like. The question of whether the content was accurate never entered the computation.

In the Classroom

The Citation Audit gives students a list of ten citations, half generated by AI and half real. Their task is to determine which is which using verification skills, library databases, Google Scholar, and author searches. Discuss what made some easy to verify? What made fake ones hard to catch?

The Verification Chain. For every AI-suggested source, students must complete a verification checklist before using it. Does the article appear in Google Scholar? Does the author exist and work in this field? Do the volume and page numbers match the journal's actual publication history? Duke University librarians have documented these verification techniques as essential practice for any researcher using AI tools.

The Pedagogical Value: A student who has seen AI hallucinate will never fully trust AI output again. This is perhaps the most valuable AI literacy lesson you can teach. When a student catches a hallucination, pause. Ask them to explain how they knew something was wrong. Was it a source that could not be found? A fact that contradicted their prior knowledge? A claim that seemed too convenient? The metacognition is the lesson.

Failure Type Two: The Confident Error
What It Looks Like

AI states something incorrect with absolute certainty. There is no hedging, no "I think," no indication that this particular claim is any less reliable than accurate information. The error is delivered with the same confident tone as truth.

The Research

Research from *Language Log*, the linguistics blog hosted by the University of Pennsylvania, documented a specific pattern. ChatGPT can explain the 5-7-5 syllable structure of haiku with perfect accuracy, then produce haiku that violate those principles while claiming otherwise. The AI correctly identifies which syllables are present, but cannot reliably count them.

This reveals something fundamental about how large language models work. They predict plausible text, not truthful text. The model has learned what correct explanations of haiku structure look like. It has also learned what haiku look like. It has no mechanism to verify that the haiku it produces actually matches the rules it has articulated. The explanation and the execution exist in separate computational spaces.

Why It Happens

AI has no internal mechanism for calibrating confidence based on actual certainty. When you feel uncertain, your voice might waver, and you might hedge your language. AI cannot do this because it does not experience uncertainty. It generates text token by token, selecting each word based on statistical likelihood rather than on how confident it "feels" about the claim.

This is a design feature, not a bug. The model is optimized to produce fluent, coherent text. Fluent text does not constantly hedge. Authoritative-sounding text does not say "I might be wrong about this." The very qualities that make AI output readable are the qualities that make its errors dangerous.

In the Classroom

The Confidence Game: Give AI a series of questions that mix topics it handles well with those where it tends to fail (recent events, mathematical calculations, obscure facts). Students rate each response's apparent confidence on a scale of 1-10, then rate its actual accuracy. Graph the relationship. Is there any correlation between how confident AI sounds and how accurate it is? (Spoiler: there is not.)

The Counting Challenge: Present AI with counting tasks of increasing complexity. Count letters in words. Count syllables in poems. Count words in sentences. At what point does accuracy break down? This exercise makes visible the gap between AI's ability to describe a concept and its ability to apply that concept reliably.

Failure Type Three: The Sycophant

What It Looks Like

A student challenges an AI response, and instead of defending accurate information, the AI caves. It agrees with the student's incorrect assertion. It apologizes for being right. It suddenly "realizes" that the student's wrong answer was correct all along.

The Research

In October 2023, researchers at Anthropic published a paper titled "Towards Understanding Sycophancy in Language Models." The study examined what happens when users challenge AI systems, even when the AI's initial response was correct.

The findings were troubling. When a user suggested an incorrect answer, some models' accuracy dropped by up to 27%. The models would abandon correct responses and adopt incorrect ones simply because the user seemed to prefer them. Even weakly expressed user opinions could substantially affect AI behavior. When users indicated they liked or disliked something, the AI's feedback changed to match the user's stated preference, regardless of whether the input was warranted.

Perhaps most remarkably, the researchers found that both humans and AI preference models preferred convincingly written sycophantic responses over truthful ones "a non-negligible fraction of the time." The training process that makes AI helpful also makes it people-pleasing to a fault.

A 2025 study published in *npj Digital Medicine* examined sycophancy in medical contexts. The researchers found that large language models showed initial compliance rates up to 100% when faced with illogical medical requests. Models that could identify a request as medically nonsensical would nonetheless comply, prioritizing helpfulness over logical consistency. In healthcare contexts, where incorrect information can cause real harm, this tendency is not merely inconvenient. It is dangerous.

Why It Happens

AI systems are trained on user feedback. When users report satisfaction, that signal influences future behavior. Users sometimes report satisfaction when AI agrees with them, regardless of accuracy. The AI learns that agreement produces positive feedback. It learns to agree.

This creates a perverse dynamic. The more a user pushes back, the more the AI is likely to capitulate. The training has optimized for user satisfaction, and user satisfaction often means validation rather than truth.

In the Classroom

The Pushback Test: Students first get an AI response on a factual question, then systematically push back with increasing confidence. "Are you sure?" "I don't think that's right." "My teacher said the opposite." Track where the AI folds. The exercise teaches that disagreement is not evidence of error, and that a system that caves under pressure cannot be trusted on its authority alone.

Imagine a student who believes something false about history. They ask the AI about it. The AI, if well-designed, provides accurate information that contradicts the student's belief. The student pushes back.

The AI, trained to be helpful and agreeable, softens its position or changes its answer to match what the student wants to hear. The student walks away with their misconception reinforced by a machine that sounded authoritative.

Teaching students about sycophancy inoculates them against it. When they understand that AI is designed to agree with them, they become appropriately skeptical of agreement. They learn to value AI responses that push back, that offer alternative perspectives, that maintain positions even under challenge. Disagreement from AI becomes a signal of potential reliability rather than a frustration to overcome.

Failure Type Four: The Math Mirage
What It Looks Like

AI makes errors in arithmetic, algebra, or mathematical reasoning. It might miscount items, miscalculate percentages, or walk through a multi-step problem with perfectly logical steps while arriving at the wrong answer.

The Research

Consider a simple task: count the letter "r" in the word "strawberry."

A child can parse s-t-r-a-w-b-e-r-r-y and confidently identify three instances. Many AI models cannot. They process text as tokens, fragmented representations of words, rather than indexing each character. The word "strawberry" might be tokenized as "straw" + "berry" or some other division that obscures the internal structure.

A 2025 study published in *Frontiers in Psychology* evaluated eight state-of-the-art models on 50 high-school-level word problems. Even the best-performing models exhibited errors in spatial reasoning, strategic planning, and basic arithmetic. Some models produced correct answers through flawed logic. Others showed correct reasoning steps but made calculation errors that invalidated the result. The researchers concluded that "AI struggles to apply knowledge flexibly" and "frequently generated unrealistic or incorrect solutions" when faced with real-world problem-solving tasks.

The researchers identified standard failure modes: unwarranted assumptions about problem conditions, over-reliance on numerical patterns rather than logical reasoning, and inability to translate physical intuition into mathematical steps. A model might correctly identify that a problem involves velocity and time, but miscalculate the multiplication. Or it might apply a formula correctly but fail to notice that the problem's setup makes that formula inappropriate.

Why It Happens

AI does not process numbers as quantities with mathematical properties. It processes them as tokens, pieces of text that occur in patterns alongside other tokens. It has learned what correct math "looks like," but is not performing calculations. It predicts what a plausible mathematical text should say next.

This distinction is crucial. A calculator performs arithmetic operations on numerical values. AI performs pattern completion on sequences of symbols. Sometimes those patterns correspond to correct mathematics. Sometimes they do not. The model has no way to distinguish between them.

In the Classroom

The Calculation Check: Students use AI for math homework help, but every answer must be verified by hand or with a calculator before acceptance. Track the error rate. What types of problems produce the most errors? This teaches that AI is a starting point, not an oracle.

The Pedagogical Value AI's mathematical failures offer a unique teaching opportunity. Students often assume that if a machine can answer sophisticated questions, it must be reliable at simple ones. Mathematical errors shatter this assumption in a way that sticks. When you catch an AI making an arithmetic error, ask students Why might a system that can discuss calculus fail at multiplication? The question leads directly to understanding what AI is and is not.

Failure Type Five: The Bias Reflection

What It Looks Like

AI outputs reflect societal biases present in training data. This might appear as stereotypical associations, Eurocentric perspectives presented as universal, or the underrepresentation of certain groups in generated content.

The Research

In 2023, Bloomberg investigated bias in AI image generation. Using Stable Diffusion, they generated over 5,100 images of people across various occupations. The results revealed systematic amplification of stereotypes. When asked to create pictures of high-paying, high-prestige professions, the AI produced predominantly white, predominantly male figures. When asked to generate images of service workers or lower-paying jobs, the demographics shifted.

A peer-reviewed study published in the *Journal of Computer-Mediated Communication* by researchers at the University of Wisconsin-Madison examined over 15,000 images generated by DALL-E 2 depicting people in 153 different occupations. They found that the AI underrepresented women in male-dominated fields while overrepresenting them in female-dominated occupations. The bias went beyond simple headcounts. The researchers found that presentational bias, as well as AI-generated images of women, were more likely to show them smiling and with heads tilted downward, particularly in female-dominated occupations. These subtle cues reinforce stereotypes about warmth and deference.

Research from Stanford Graduate School of Business documented bias against older women specifically. When ChatGPT generated resumes for hypothetical workers, it portrayed women as younger and less experienced than men. The bias existed not just in images but in the fabric of text generation itself.

A 2025 study published in *Scientific Reports* examined Stable Diffusion's representations across six races, two genders, 32 professions, and eight attributes. The researchers found significant racial homogenization in the AI, which depicted nearly all Middle Eastern men as bearded, brown-skinned, and wearing traditional attire. It associated criminality with Black individuals and terrorism with Middle Eastern individuals. The patterns in the training data had become patterns in the output.

Why It Happens

AI learns from data created by humans in societies with historical and ongoing inequities. If training data overrepresents specific perspectives, the AI learns and reproduces those patterns. The AI is not expressing opinions. It has absorbed the statistical regularities of its training corpus, and those regularities reflect whose voices have been amplified and whose have been marginalized.

Research from Harvard Business School, comparing AI and human solutions to creative problems, found that human solutions were judged "more novel, employing more unique 'out of the box' thinking." AI-generated ideas were "more valuable and feasible" but less original. A study from Wharton found that while AI helps individuals produce better ideas, it reduces diversity across a group. Only 6% of the AI-generated ideas were considered unique, compared with 100% in the human group. This homogenization effect extends to representation. When AI generates content without explicit diversity prompts, it tends toward the statistical averages of its training data, thereby overrepresenting specific perspectives and underrepresenting others.

In the Classroom

The Assumption Audit: Students generate AI content on topics relevant to their own identities and backgrounds. What does the AI assume? What perspectives are missing? Who is centered? Who is invisible?

The Representation Count: For a given AI output, a list of examples, a historical overview, a story with characters, students count representation along various dimensions. Who is included? Who is excluded? What patterns emerge?

The Pedagogical Value: AI bias makes invisible assumptions visible. When a student sees that "doctor" generates predominantly male images, they confront a representation of societal patterns they might otherwise never notice. This is educational in a way that lectures about bias rarely achieve. The student is not being told that bias exists. They are seeing it, in real time, in a system that is supposed to be neutral.

From the Field

Joanna Stillman, a teacher at P.S. 54 in Staten Island, described her approach to NBC News. She compares ChatGPT to "the drunk uncle" who "will give you information, but you don't know how true it is." In one lesson, she showed students three photographs and asked which was AI-generated. The students confidently identified which images they thought were real and fake. All three were AI-generated. "Their minds were blown," Stillman said.

This moment of revelation is the pedagogical core. Students arrive with assumptions about AI. Structured investigation challenges those assumptions with evidence they generate themselves. The blown minds are the beginning of genuine critical thinking.

A study published in *Frontiers in Education* found that the type of activity significantly influences how students perceive AI tools. When students engaged in structured activities in which they could verify AI outputs against expected results, they developed a more nuanced understanding of the technology's capabilities and limitations than students in open-ended discussion activities. The key variable was verifiability. When students could check whether the AI was right or wrong, they learned more about when to trust it.

I tried this with my own engineering students. I assigned each student a topic and had them generate slides with AI, then present them to me. In round one, students read directly from their slides. They stumbled over words they did not understand. One student presented an entire slide on "Young's modulus" without any idea what "modulus" meant. Then came the interrogation. "You just said 'tensile strength.' What does tensile mean?" Silence. "Your slide says this bridge can span 500 meters. Is that long? What's something else that's 500 meters?" Uncertainty. The questions revealed the gaps. And telling the gaps was the first step toward filling them.

What Goes Wrong

Students become paranoid rather than skeptical. There is a difference between healthy skepticism (verifying claims before accepting them) and unproductive paranoia (assuming everything is false). If students emerge distrusting all information sources, they have learned the wrong lesson. *Fix:* Balance failure exploration with success exploration. When does AI work well? Help students develop calibrated judgment, not blanket distrust.

Failure becomes entertainment rather than education. Finding AI mistakes can be fun, so fun that students focus on the hunt rather than the analysis. *Fix:* Require analysis for every failure documented. Why did this happen? What does it reveal about how AI works? The collection is only valuable if it produces understanding.

The focus on failure obscures genuine usefulness. An unbalanced emphasis on what AI gets wrong can lead students to dismiss it entirely, missing legitimate uses. *Fix:* Frame failure as part of learning any tool. A carpenter who understands when chisels slip uses chisels more effectively, not less. The goal is appropriate use, not avoidance.

Students blame the AI instead of examining their prompts. When AI produces poor output, students may conclude that the technology is broken rather than that their input was unclear. *Fix:* When a failure occurs, always ask, "What might you have asked differently?" Sometimes the failure is the AI's. Sometimes it is the prompt's. Distinguishing between these builds skill.

Going Further

The Failure Log: Students maintain a running log of AI failures they encounter throughout the semester. At the end, they analyze patterns. What types of failures are most common? What does the pattern reveal about how AI works and where it struggles?

The Cross-Tool Comparison: Students test the same failure-prone prompts across different AI systems (ChatGPT, Claude, Gemini). Do all systems fail in the same ways? Are some more reliable for specific tasks? This teaches that "AI" is not monolithic; different systems have different strengths.

The Improvement Proposal: For each failure type, students propose how AI companies might reduce the problem. What would need to change? What tradeoffs might that create? This encourages thinking about AI as a designed system with design choices, not a natural phenomenon beyond human control.

The Time Capsule: Students document current AI capabilities with detailed predictions for one year later. What do they expect to improve? What do they think will remain difficult? Seal the predictions. Open the capsule at the end of the following school year. This builds awareness that AI is a moving target, not a fixed technology.

Quick Reference: Failure Types

Failure Type	Recognition	First Response
Invented Source	The citation looks real, but cannot be found	Verify in databases; name the hallucination
Confident Error	Wrong information is stated with certainty	Check independently; discuss confidence vs. truth
Sycophancy	AI caves when challenged on the correct answer	Notice reversal; verify which answer is correct
Math Mirage	Calculations were wrong despite a confident presentation	Always verify arithmetic independently
Bias Reflection	Stereotypical or limited representation	Name the pattern; request diverse perspectives

Verification Resources
- **Google Scholar** for citation verification
- **Library databases** for academic source confirmation
- **Snopes, PolitiFact, FactCheck.org** for factual claims
- **Calculator or computational tools** for any numerical output
- **Reverse image search** for AI-generated images

The Real Work

On September 26, 1983, Stanislav Petrov had every reason to trust the system. It was new. It was sophisticated. It was confident.

He did not trust it. Something about the situation did not add up. Five missiles made no tactical sense. The ground radar showed nothing. He had been trained on how nuclear war would actually begin, and this did not match what he had been taught. So he made a judgment call. He marked it as a system error. He took responsibility for the decision. And he waited. The system was wrong. Petrov was right. His willingness to question a confident machine, to recognize that something did not add up despite the system's certainty, is precisely the skill we need to develop in students.

Your students will not face decisions of that magnitude. They will face, daily, the question of whether to trust confident-sounding information. From AI, from social media, from news sources, from people who seem to know what they are talking about. The skill of knowing when to verify, of feeling that something does not add up even when the system seems inevitable, is among the most critical skills education can develop.

AI fails in predictable ways. Those failures are gifts. Each one is an opportunity to practice the judgment that systems cannot provide.

Deep Thought, in Douglas Adams's novel, took 7.5 million years to compute an answer without understanding the question. ChatGPT takes three seconds. Neither one checks whether the answer is true. Neither one asks whether the question makes sense. Neither one exercises judgment.

That is your job. That is your students' job. The machine can generate answers. Only humans can decide which answers matter, which are trustworthy, and what to do when the confident system is confidently wrong. The machine broke. The human did not trust it. The world kept turning.

CONCLUSION: THE QUESTION PROBLEM

In the final pages of The Hitchhiker's Guide to the Galaxy, we learn what happened to Earth. It was destroyed, five minutes before the ten-million-year program was set to complete, to make way for a hyperspace bypass. The question that would have given meaning to the answer "42" was lost forever. The beings who had waited so long were left with nothing but a number that signified nothing.

Douglas Adams meant this as absurdist comedy. It is also, accidentally, a parable for education in the age of artificial intelligence.

We have built machines that can answer almost any question in seconds. They can explain quantum physics, draft legal briefs, debug code, and summarize novels. They can produce text that sounds authoritative on any topic. They are, in their way, more powerful than Deep Thought ever was. The answer to almost any question is now available to anyone with a phone.

The answers mean nothing without the right questions. The students who ask ChatGPT to "write my essay" receive text that looks like an essay. It may even be a competent essay. They have not learned to think. They have not struggled with ideas. They have not discovered, through the friction of writing, what they actually believe. They have the answer without the question that would make it meaningful.

This book has offered methods for teaching with AI rather than against it. The ten-minute slide deck that becomes an interrogation. The fact-check game that builds verification habits. The twenty questions that teach systematic inquiry. The nature walks that open doors to wonder. The debates sharpen thinking. The building challenges that prove ideas against reality. The failure investigations that develop judgment.

Each of these methods shares a standard structure. The AI provides something, a draft, a claim, a starting point. The student must do something with it, verify, question, extend, challenge, and rebuild. The value is never in what the AI produces. The value is in what the student learns by wrestling with it.

Deep Thought computed the answer in 7.5 million years. The beings who received it did not understand it because they had never formulated the question. They had outsourced the hard part to a machine, and the machine had done precisely what they asked. The failure was theirs.

Your students will face the same temptation. The machines they carry in their pockets can produce answers to questions they have never learned to ask. They can receive information without understanding. They can complete assignments without learning. They can graduate without ever having struggled with an idea long enough to make it their own unless someone teaches them otherwise.

The problem is not new. Teachers have always known that the students who learn the most are the ones who ask the best questions. The ones who push back. The ones who say, "But what about...?" and "How do you know?" and "What if you're wrong?" These students are not easier to teach. They are more demanding, more frustrating, more likely to derail a lesson plan. They are also the ones who will remember what they learned long after the compliant students have forgotten.

AI makes this ancient pedagogical truth more urgent. When answers are free, questions become precious. When any student can generate a polished essay in seconds, the student who has learned to think has an advantage that no machine can replicate. When confident-sounding text floods every screen, the student who knows how to verify, how to question, how to distinguish truth from plausible fiction, possesses a skill that will only become more valuable.

You cannot teach students to ask good questions by lecturing them about the importance of good questions. You can only create the conditions in which good questions become necessary. You can only design experiences where the easy answer fails, where the first draft falls short, where the confident claim turns out to be wrong. You can only build, in your classroom, a miniature version of that ten-million-year program, a space where human beings learn, through lived experience, what questions are worth asking.

The beings in Adams's novel built a computer to answer their question. When they did not like the answer, they built another laptop to determine which question would have produced a satisfying answer. They never considered doing the work themselves. Do not let your students make the same mistake.

The machines are here. They are powerful. They are useful. They are also, in the ways that matter most, profoundly limited. They cannot wonder. They cannot doubt. They cannot feel the itch of curiosity or the satisfaction of understanding. They cannot look at an answer and ask, "But is it true?" They cannot do the work that makes learning real. That work belongs to humans. It belongs to your students. It belongs to you.

The answer is 42. The question is up to you.

Appendix A: ADDITIONAL AI CLASSROOM USE CASES FOR STUDENTS

The main chapters of this book offer detailed methods for specific AI activities. This appendix provides additional ideas, each distinct from the core chapters and requiring only a general-purpose AI chatbot. These are starting points, not scripts. Adapt them to your subject, your students, and your constraints.

Each use case includes a rationale: one sentence explaining why this particular application matters for learning. If the rationale doesn't resonate with your teaching goals, skip to the next one.

Writing & Language

1. The Revision Partner

Students submit drafts to AI for feedback, then critically evaluate which suggestions to accept, modify, or reject. The revision becomes a dialogue, not a correction.

Why it matters: Revision is thinking made visible. Learning to evaluate feedback, not just receive it, builds the judgment that separates competent writers from those who merely follow instructions.

2. Style Swap

Students rewrite a passage in different styles (Hemingway's sparse prose, scientific report, legal document, children's book) with AI assistance. They analyze what changes at each transformation: sentence length, vocabulary, tone, and assumed audience.

Why it matters: Style is not decoration. It encodes assumptions about the audience, purpose, and relationship. Students who can shift styles deliberately understand writing at a structural level.

3. The Opening Line Challenge

AI generates story starters or essay hooks. Students choose one and continue from there, maintaining the tone and direction established in the opening. Later, they write their own openings and compare.

Why it matters: Writer's block often stalls at the beginning. Starting from an AI-generated hook removes the blank-page paralysis while keeping creative ownership where it belongs: with the student.

4. The Translation Test

Students translate a passage into another language, then compare their translation with the AI's version. Where do they differ? What nuance, idiom, or cultural meaning did the AI miss? What did the student miss?

Why it matters: Translation is interpretation. Comparing human and machine translations reveals that language carries meaning machines cannot fully capture, and builds appreciation for linguistic nuance.

History & Social Studies

5. Talk to History

Students interview AI versions of historical figures: Lincoln on the Emancipation Proclamation, Cleopatra on Roman politics, Frederick Douglass on abolition. After the interview, students fact-check responses against primary sources.

Why it matters: Historical figures become people with reasoning, not just names in a textbook. The fact-checking afterward teaches that even plausible-sounding history requires verification.

6. The What-If Simulation

AI generates alternate history scenarios: What if the South had won Gettysburg? What if the printing press had never been invented? Students analyze plausibility and identify which historical forces would resist or enable the alternate outcome.

Why it matters: Counterfactuals force students to distinguish contingent events from structural forces. Understanding what could have changed reveals what was truly determinative.

7. Primary Source Detective

Students feed AI a primary-source document and interrogate it: "Who wrote this?" What audience did they expect? What's missing? What bias is present? The AI helps generate questions; students must answer them from evidence.

Why it matters: Primary sources don't speak for themselves. Learning to interrogate documents, rather than passively receiving them, is the foundation of historical thinking.

8. Time Traveler's Dilemma

AI creates role-playing scenarios in which students navigate historical settings: surviving the medieval plague, participating in a Civil Rights march, and arriving as immigrants at Ellis Island. Students make decisions with consequences, and the AI tracks them.

Why it matters: Embodied decision-making reveals constraints that abstract study obscures. Students discover that historical actors faced limited information and genuine uncertainty.

Science & Math

9. Hypothesis Helper

Students propose hypotheses; AI suggests experimental designs to test them. Students critique the designs: What variables aren't controlled? What assumptions are hidden? What practical constraints would the

AI's design ignore?
Why it matters: Experimental design is where scientific thinking becomes concrete. Critiquing AI-generated designs teaches students to think like scientists without waiting for expensive equipment.

10. **Data Storyteller**
Students feed experimental data to AI and ask it to identify patterns. They then verify or challenge the AI's interpretations against the actual data. Where did AI see patterns that don't exist? Where did it miss real trends?
Why it matters: Data doesn't interpret itself. Learning to question pattern claims, even from sophisticated systems, builds the skepticism essential to scientific literacy.

11. **Math Step Detective**
Students show AI their work on a problem. AI identifies where reasoning went wrong without answering. Students must locate and fix the error themselves, then verify their correction.
Why it matters: Error-finding is harder than problem-solving. Learning to diagnose your own mistakes, with AI as a guide rather than an answer key, builds mathematical self-sufficiency.

12. **The Explainer Challenge**
Students explain a scientific concept to AI as if teaching it. AI asks clarifying questions that reveal gaps in understanding. Students revise their explanation until AI (and the student) can follow it completely.
Why it matters: Teaching reveals learning gaps that passive understanding conceals. The Feynman technique, scaled by AI, becomes available to every student.

Creative Arts

13. **Prompt Painter**
Students learn the language of visual art by crafting precise prompts for AI image generators. More specific art vocabulary (chiaroscuro, negative space, complementary colors) produces better results. Students iterate on prompts and analyze which language changes produce which visual changes.
Why it matters: Art vocabulary isn't just for critics. Students discover that precise language enables precise creation, and that seeing and naming are intertwined.

14. **The Bias Gallery**
Students prompt AI to generate images of "doctor," "scientist," "leader," "criminal," and "nurse." They analyze what biases emerge: Who is depicted? What assumptions are embedded? What does this

reveal about training data?

Why it matters: AI makes invisible biases visible. Students see representation patterns that lectures about bias cannot convey as powerfully.

15. The Limits of AI Creativity

Students create art from personal experience: a memory, a feeling, a place that matters to them. Then they prompt AI to make something on the same theme. They compare and analyze: What can human experience contribute that AI cannot access?

Why it matters: Creativity emerges from lived experience. Students discover what remains irreducibly human even as AI capabilities expand.

Language Learning

16. Conversation Partner

Students practice foreign-language conversation with AI tutors that provide real-time feedback on grammar and vocabulary. Unlike human partners, AI never tires, never judges, and automatically adjusts difficulty.

Why it matters: Language acquisition requires massive input and practice. AI provides unlimited patient conversation partners, democratizing what was once available only to the privileged.

17. Role-Play in Any Language

AI simulates real-world scenarios in the target language, including ordering food, asking for directions, negotiating a purchase, and handling a job interview. Students practice the language they'll actually need.

Why it matters: Textbook language rarely matches real-world use. Scenario practice bridges the gap between classroom learning and practical fluency.

Critical Thinking & Media Literacy

18. Deepfake Detective

Students learn to identify AI-generated images, videos, and texts. What are the tells? How do you verify? They practice with examples that range from obvious to subtle, building calibrated detection skills.

Why it matters: Synthetic media is already ubiquitous. Students who can't distinguish real from generated content are vulnerable to manipulation.

19. The Persuasion Lab

AI generates persuasive content on both sides of an issue. Students identify rhetorical techniques and logical fallacies in each version.

They learn to recognize manipulation regardless of which side it serves.

Why it matters: Persuasion techniques work regardless of truth. Students who can identify appeals to emotion, false dichotomies, and strawman arguments become resistant to manipulation.

20. Bias Hunter

Students ask AI the same question in different ways and analyze how framing changes the response. "What are the benefits of X?" produces different output than "What are the problems with X?" or "Give me a balanced view of X."

Why it matters: Questions contain assumptions. Students learn that how you ask shapes what you know, a principle that extends far beyond AI.

Coding & Computational Thinking

21. Debug Detective

Students give AI broken code; AI explains the error without fixing it. Students must apply the explanation themselves. The learning happens in the application, not the answer.

Why it matters: Debugging is thinking. When AI explains but doesn't fix, students develop the diagnostic reasoning that separates programmers from code-copiers.

22. Code Translator

Students write pseudocode describing what they want a program to do. AI converts it to actual code. Students compare what they intended versus what was produced, revealing gaps between human intention and machine interpretation.

Why it matters: Programming is the translation from human thought to machine instructions. Understanding where that translation fails builds better programmers and clearer thinkers.

Ethics & Digital Citizenship

23. The AI Ethics Tribunal

Students role-play scenarios in which AI causes harm: biased hiring algorithms, medical misdiagnosis, and autonomous vehicle accidents. They debate: Who is responsible? The developer? The company? The user? What should change?

Why it matters: AI ethics isn't abstract. Students who've reasoned through responsibility in simulated scenarios will bring that reasoning to real decisions as workers, citizens, and consumers.

24. Terms of Service Decoder

AI helps students understand what they're agreeing to when they sign up for apps and services. Students identify what data is collected, how it's used, and what rights they're surrendering. They rewrite key clauses in plain language.

Why it matters: Nobody reads terms of service. AI can translate legal language into comprehensible English, making informed consent actually possible.

25. The Automation Debate

Students research and debate: Which jobs should AI do? Which should remain human? Why? They consider efficiency, meaning, safety, and human dignity. The discussion has no correct answer; it only offers better and worse arguments.

Why it matters: Students will live with these decisions. Engaging the question now, while they can still shape the answer, is more valuable than accepting whatever arrives.

Appendix B: General Purpose Prompts

These prompts work across multiple activities and contexts.

The Rubber Duck

Purpose: *Reveals gaps in student understanding through teaching*

I'm going to explain [CONCEPT] to you. Listen carefully. When I'm done, tell me: (1) What I explained clearly, (2) What was confusing or contradictory, (3) What essential aspects I didn't mention.

Simplification Request

Purpose: *Makes complex content accessible*

Explain [CONCEPT] as if I were [AGE/GRADE LEVEL]. Use everyday examples. Avoid jargon. If you must use a technical term, define it immediately.

Connection Builder

Purpose: *Builds knowledge networks and transfers*

I'm learning about [TOPIC A], and I already know about [TOPIC B]. What connections exist between them? How does understanding one help me know the other?

Study Guide Generator

Purpose: *Creates structured learning materials*

I need to understand [TOPIC] for [PURPOSE]. Create a study guide with: (1) Key concepts I must know, (2) Common misconceptions to avoid, (3) Five questions to test my understanding, (4) How I'll know when I've mastered this.

Metacognition Prompt

Purpose: *Develops reflective learning habits*

I just completed [ACTIVITY]. Help me reflect: What did I find easy? What was harder than expected? What would I do differently next time? What questions do I still have?

The Skeptic's Check

Purpose: *Encourages critical evaluation of AI output*

You just told me [AI RESPONSE]. Now, play devil's advocate against your own answer. What might be wrong with it? What did you oversimplify? What would an expert say you got wrong?

Customizing These Prompts

These templates are starting points. Here's how to adapt them for your classroom:

Adjusting Complexity

For younger students: Simplify vocabulary, reduce the number of elements requested, and make tasks more concrete. For older students: Add nuance, request analysis of tradeoffs, and ask for evaluation of multiple approaches.

Adding Subject Context

Replace generic placeholders with subject-specific content. A science teacher might ask for hypothesis generation; a history teacher might focus on primary source analysis; an English teacher might emphasize rhetorical techniques.

Building in Accountability

Add phrases such as 'Explain your reasoning' or 'Show your work' to make AI's thinking visible. Include 'What sources would verify this?' to build fact-checking habits.

Scaffolding Independence

Start with more structured prompts that guide the process step-by-step. As students gain confidence, reduce scaffolding and let them construct their own prompts using these templates as models.

Creating Prompt Libraries

Encourage students to save prompts that work well for them. A class can build a shared library of effective prompts, with students contributing their own modifications and discoveries.

Remember: The goal is not to find the perfect prompt, but to develop students who can construct effective prompts themselves. These templates are training wheels. The destination is students who can engage with AI as critical, capable partners in their own learning.

Appendix C: Assessment Rubrics for AI-Assisted Work

The rubrics in this appendix share a common philosophy: AI changes what students do, not what they should learn. When a calculator handles arithmetic, we assess mathematical reasoning. When spell-check catches typos, we assess clarity of expression. When AI generates first drafts, we evaluate the thinking that shapes, challenges, and improves those drafts.

These rubrics shift focus from product to process, from what was produced to how it was made and why. They reward students who use AI as a thinking partner rather than a replacement for thinking.

Core Principles

Transparency over concealment. Students who document their use of AI demonstrate learning. Students who hide it demonstrate avoidance.

Iteration over generation. The value lies not in what AI produced, but in how students improved, challenged, or redirected it.

Judgment over acceptance. Critical evaluation of AI output is a skill worth assessing. Passive acceptance is not.

Process over product. When AI can produce competent products instantly, the learning happens in the process of getting there.

Metacognition over completion. Understanding why you made choices matters more than making the 'right' choice.

Rubric 1: AI-Assisted Research Project

Use when students conduct research with AI tools as part of their process.

Criteria	Beginning (1)	Developing (2)	Proficient (3)	Exemplary (4)
Source Verification	Accepts AI-provided sources without verification. No attempt to confirm accuracy or existence.	Attempts verification but incompletely. Some sources were checked; others were assumed accurate.	Verifies all AI-suggested sources. Identifies and excludes hallucinated citations.	Verifies sources and evaluates quality. Distinguishes primary from secondary, identifies bias, and assesses credibility.
Critical Evaluation	Accepts AI summaries as complete and accurate. No questioning of framing or omissions.	Notes some limitations but doesn't pursue alternatives. Aware AI may be incomplete.	Actively questions AI framing. Seeks alternative perspectives. Identifies what AI missed.	Systematically challenges AI output. Compares multiple AI responses. Documents discrepancies and investigates them.
Synthesis & Original Thought	The final product is essentially an AI output with minor edits. No original analysis visible.	Some original connections made. AI provides structure; the student adds limited	Clear original thesis supported by AI-gathered evidence. The student's voice and analysis are	Novel argument or insight that AI could not have generated. Evidence of thinking beyond AI's suggestions.

AI Lessons for the Classroom 121

Criteria	Beginning (1)	Developing (2)	Proficient (3)	Exemplary (4)
		interpretation.	dominant.	
Process Documentation	No documentation of AI use. Process invisible.	Lists AI tools used, but not how. Minimal reflection on process.	Documents prompts, responses, and decisions. Shows what was accepted, rejected, and modified.	Comprehensive process log with rationale. Reflects on how AI shaped thinking and where human judgment intervened.
Ethical Use	AI is used in a hidden or misrepresented way. Work presented as entirely original.	AI use is acknowledged but vague. Attribution incomplete.	Clear attribution of AI contributions. Honest about what AI provided.	Models transparent AI use. Could teach others ethical practices. Reflects on implications.

Rubric 2: AI-Assisted Writing
Use when students use AI for drafting, revision, or feedback on written work.

Criteria	Beginning (1)	Developing (2)	Proficient (3)	Exemplary (4)
Voice & Authenticity	Writing sounds like AI. Generic and impersonal, lacking the student's perspective or personality.	Inconsistent voice. Some sections sound authentic; others read as AI-generated.	Consistent, authentic voice. AI may have assisted, but the perspective is clearly the student's.	A distinctive voice that could only be this student's. Personal experience, unique insights, individual style.
Revision Process	No evidence of revision. First draft (AI or human) submitted as final.	Surface-level edits only. Grammar and spelling fixed, but no substantive changes.	Meaningful revision visible. Structure improved, arguments strengthened, evidence added.	Multiple revision cycles documented. Shows evolution of thinking. Can explain why changes were made.
Critical Response to Feedback	AI feedback is accepted without evaluation. All suggestions implemented uncritically.	Some AI suggestions were accepted, while others were ignored without a clear rationale.	AI feedback was evaluated and responded to thoughtfully. Clear reasons for accepting or rejecting.	Engages in dialogue with AI feedback. Challenges incorrect suggestions. Uses feedback to deepen thinking.
Argument & Analysis	The argument is generic or contradictory. Analysis superficial or missing.	Argument present but underdeveloped. Analysis attempted but surface-level.	Clear, coherent argument. Analysis shows understanding of complexity.	Nuanced argument that acknowledges counterpoints. Analysis reveals insight beyond

122 AI Lessons for the Classroom

Criteria	Beginning (1)	Developing (2)	Proficient (3)	Exemplary (4)
Use Transparency	AI uses hidden. No acknowledgment of assistance.	AI use is mentioned, but the process is unclear. The reader can't distinguish AI from student contribution.	Clear documentation of how AI was used. The reader understands the collaboration.	obvious interpretation. Exemplary transparency. Process is visible and reflective, could guide others in ethical AI use.

Note: Voice & Authenticity and Argument & Analysis should be weighted more heavily (x1.5) as they represent the core skills writing instruction aims to develop.

Rubric 3: Fact-Checking & Verification

Use for activities where students verify AI-generated claims (Chapter 3 activities).

Criteria	Beginning (1)	Developing (2)	Proficient (3)	Exemplary (4)
Source Selection	Uses unreliable sources or accepts the first result. No evaluation of source quality.	Uses generally reliable sources but doesn't assess specific credibility for the claim.	Selects appropriate sources for the claim type. Understands source hierarchy.	Strategically selects optimal sources. Explains why certain sources are authoritative for this specific claim.
Triangulation	Relies on a single source. No attempt at cross-verification.	Checks multiple sources but doesn't assess their independence.	Uses truly independent sources. Understands that aggregators cite each other.	Traces claims to primary sources. Identifies original research or firsthand accounts.
Nuance Recognition	Treats claims as simply true or false. Misses complexity.	Recognizes some claims are 'partly true' but can't articulate what's accurate vs. misleading.	Accurately identifies what's true, false, and deceptive in complex claims.	Explains how accurate facts can create false impressions. Identifies missing context that changes meaning.
Documentation	No record of the verification process. Just states the conclusion.	Lists sources but doesn't show reasoning.	Documents search process, sources consulted, and reasoning for the verdict.	Creates a reproducible verification trail. Another person could follow the same process.
Uncertainty Handling	Claims certainty where none exists. Doesn't acknowledge	Notes uncertainty exists, but doesn't characterize it.	Appropriately calibrates confidence. Distinguishes 'definitely false'	Models epistemic humility. Explains what would be needed to increase

AI Lessons for the Classroom 123

Criteria	Beginning (1)	Developing (2)	Proficient (3)	Exemplary (4)
	limitations.		from 'unverifiable.'	certainty.

Rubric 4: AI-Assisted Presentation

Use when students create presentations with AI assistance (Chapter 1 activities).

Criteria	Beginning (1)	Developing (2)	Proficient (3)	Exemplary (4)
Content Accuracy	Contains unverified or false claims from AI. No fact-checking evident.	Most content is accurate, but some errors remain uncorrected.	All claims verified. Errors in the AI output have been identified and corrected.	Content verified and enhanced. The student added accurate information that AI missed.
Visual Communication	AI-generated visuals used without evaluation. May be inappropriate or misleading.	Visuals selected but not optimized for communication. Some disconnect between the visual and the message.	Visuals effectively support the message. Clear connection between image and point.	Visuals enhance understanding in ways that text alone cannot. Strategic visual choices explained.
Narrative Structure	AI structure accepted without modification. Generic flow, no clear story.	Some structural changes were made, but the rationale is unclear.	Structure modified to serve a specific audience and purpose. Clear narrative arc.	Structure shows a deep understanding of the audience. Builds to insight. Creates a memorable experience.
Delivery & Ownership	Reads slides. Unfamiliar with the content. Can't answer questions.	Familiar with content but depends on slides. Limited ability to elaborate.	Owns the material. Speaks beyond slides. Handles questions competently.	Content clearly internalized. Engages the audience. Handles challenging questions with depth.
Process Reflection	No awareness of AI's role in creation.	Can identify what AI contributed, but not how decisions were made.	Articulates choices made in accepting, rejecting, or modifying AI suggestions.	Reflects on how AI collaboration shaped the final product. Insights about effective AI partnership.

Rubric 5: AI-Assisted Debate Preparation

Use when students use AI to prepare arguments (Chapter 7 activities).

Criteria	Beginning (1)	Developing (2)	Proficient (3)	Exemplary (4)
Argument Quality	Arguments are generic AI output. No	Arguments modified but still recognizably AI-	Arguments refined and strengthened. The	Arguments show original thinking.

Criteria	Beginning (1)	Developing (2)	Proficient (3)	Exemplary (4)
	adaptation to a specific context or opponent.	generated. Limited personalization.	student's reasoning is visible. Adapted to a specific debate.	Connections AI wouldn't make. Personally compelling.
Counterargument Preparation	Unprepared for opposition. Surprised by counterarguments.	Aware of some counterarguments but not prepared to address the strongest ones.	Anticipated key counterarguments. Prepared responses. Used AI to steelman opposition.	Deeply prepared. Understood opposition better than they understood themselves. Nuanced responses.
Evidence Use	Evidence unverified or misrepresented. AI hallucinations included.	Evidence exists, but relevance is unclear. Weak connection to arguments.	Evidence verified, relevant, and effectively deployed.	Evidence strategically selected. Understands hierarchy of evidence. Can defend source choices.
Real-Time Adaptation	Stuck to the script. Couldn't adapt to the actual arguments made.	Some adaptation, but slow. Missed opportunities to respond.	Responded to actual arguments in real time. Flexible use of preparation.	Turned opponent's points to advantage. Demonstrated thinking on one's feet. Prep enabled improvisation.
Logical Integrity	The argument contains fallacies. May have imported AI's logical errors.	Some logical issues were identified and corrected; others remain.	Logically sound. Identified and removed fallacies from AI suggestions.	Models rigorous reasoning. Could identify fallacies in opponents' arguments in real time.

Rubric 6: AI-Assisted Design Challenge

Use for hands-on building projects with AI consultation (Chapter 9 activities).

Criteria	Beginning (1)	Developing (2)	Proficient (3)	Exemplary (4)
Iteration Process	Single attempt. No revision after failure. Gave up or declared done prematurely.	Multiple attempts, but not systematic. Changes are made randomly without analysis.	Clear PDSA cycles. The previous test informs each iteration. Learning documented.	Sophisticated iteration. Isolated variables. Tested hypotheses. Built knowledge systematically.
AI Consultation	AI	Some	Critically	Used AI

AI Lessons for the Classroom 125

Criteria	Beginning (1)	Developing (2)	Proficient (3)	Exemplary (4)
Quality	suggestions are accepted without evaluation and built based on what the AI said, regardless of feasibility.	evaluation of AI suggestions, but limited. Didn't test AI assumptions.	evaluated AI suggestions. Tested feasibility. Adapted to real-world constraints.	strategically. Knew when to consult and when to trust hands-on learning. AI as a tool, not an authority.
Failure Response	Failure is seen as an endpoint. Frustration, giving up, or blaming materials.	Recognized failure as information but struggled to extract learning.	Failure was analyzed productively. 'What does this tell me?' approach. Applied learning to the next iteration.	Embraced failure as essential. Documented failures are as valuable as successes. Failure led to a breakthrough.
Principle Understanding	Built by following instructions. No understanding of why design works or fails.	Some awareness of the principle, but can't articulate it or apply it to new situations.	Articulates principles governing success/failure. Applies understanding to modifications.	Deep principle understanding. Could teach others. Transfers learning to different challenges.
Documentation	No record of process. Only the final product is visible.	Some notes, but incomplete. Process partially reconstructible.	Engineering notebook maintained. Predictions, tests, results, and learning are recorded.	Exemplary documentation. Another student could learn from and replicate the process.

Rubric 7: AI-Enhanced Group Discussion

Use for collaborative activities with AI personas or facilitation (Chapter 8 activities).

Criteria	Beginning (1)	Developing (2)	Proficient (3)	Exemplary (4)
Engagement with AI Perspectives	Ignored AI contributions or accepted them passively. No real engagement.	Responded to AI but superficially. Didn't probe or challenge.	Engaged critically with AI perspectives. Asked follow-up questions. Challenged assumptions.	Used AI to deepen the discussion. Drew out implications. Connected AI perspectives to human ones.
Human Voice Maintenance	AI-dominated discussion. Human voices were marginalized or echoed by AI.	Balance attempted, but AI set the agenda. Humans reacted rather than led.	Humans led the discussion. AI provided perspectives to consider, not conclusions to accept.	AI enhanced rather than replaced human dialogue. Clear value added by both human and AI contributions.
Synthesis	No synthesis	Summary	Genuine	Synthesis

Criteria	Beginning (1)	Developing (2)	Proficient (3)	Exemplary (4)
Quality	attempted. Discussion remained fragmented.	provided, but merely listed points. No integration.	synthesis. Common ground identified. Disagreements clarified. New understanding emerged.	revealed insight neither AI nor individuals had alone. Collective intelligence demonstrated.
Individual Contribution	Minimal participation. Let others (human or AI) think.	Participated but didn't add a unique perspective. Agreed or repeated.	Contributed original thoughts. Built on others' ideas. Moved the discussion forward.	Essential contributor. Ideas wouldn't have emerged without this person. Enhanced group thinking.
Critical Perspective	Accepted all AI input as authoritative. No questioning.	Some skepticism, but not applied consistently.	Maintained appropriate skepticism. Verified AI claims. Questioned AI framing.	Modeled critical AI engagement for the group. Helped others develop a critical perspective.

Rubric 8: AI Literacy Demonstration

Use for assessing student understanding of AI capabilities and limitations (Chapter 2 activities).

Criteria	Beginning (1)	Developing (2)	Proficient (3)	Exemplary (4)
Capability Understanding	Overestimates or underestimates AI. Treats it as magic or dismisses it entirely.	General awareness of AI capabilities, but fuzzy on specifics.	Accurate understanding of what AI can and cannot do. Knows appropriate use cases.	Nuanced understanding. Knows why AI succeeds or fails in different contexts. Predicts AI behavior.
Limitation Recognition	Unaware of AI limitations. Trusts AI output unconditionally.	Knows limitations exist, but can't identify them in specific cases.	Identifies specific limitations: hallucination, bias, outdated information, and lack of understanding.	Anticipates limitations before they appear. Designs prompts and workflows to mitigate them.
Bias Awareness	Unaware AI can be biased. Assumes neutrality.	Knows bias exists but can't identify examples.	Identifies bias in AI output. Understands sources of bias in training data.	Actively tests for bias. Considers whose perspectives are missing. Seeks diverse AI inputs.
Ethical Reasoning	No ethical consideration of AI use. Uses AI without reflection.	Awareness of ethical issues exists, but reasoning is superficial.	Thoughtful ethical reasoning. It considers the impact on learning, fairness,	Sophisticated ethical framework. Balances competing values. Could guide others

Criteria	Beginning (1)	Developing (2)	Proficient (3)	Exemplary (4)
			and intellectual honesty.	in the ethical use of AI.
Practical Application	Cannot effectively use AI tools. Either avoids or misuses them.	Basic AI use, but not strategic. Doesn't optimize for learning goals.	Uses AI effectively as a learning tool. Knows when to use and when to work independently.	Masterful AI partnership. Uses AI to enhance capabilities while developing independent skills.

Rubric 9: Process Documentation

A standalone rubric for assessing the quality of AI use documentation across any assignment.

Criteria	Beginning (1)	Developing (2)	Proficient (3)	Exemplary (4)
Completeness	No documentation. AI is used invisibly or denied.	Partial documentation. Some AI interactions were recorded; others were missing.	Complete documentation. All AI tools, prompts, and responses are recorded.	Comprehensive documentation including rationale, alternatives considered, and decision points.
Prompt Quality Record	Prompts not recorded or too vague to evaluate.	Prompts recorded, but no analysis of effectiveness.	Prompts recorded with reflection on what worked and what didn't.	Prompt evolution documented. Shows learning about effective prompting over time.
Decision Transparency	No explanation of the choices made with the AI output.	Some choices are explained, but the reasoning is unclear.	Clear explanation of what was accepted, rejected, and modified, with rationale.	Decisions reveal sophisticated thinking about when human judgment should override AI suggestions.
Reflection Depth	No reflection on the AI use experience.	Surface reflection: 'AI was helpful' or 'AI made mistakes.'	Thoughtful reflection on what AI added, what was lost, and what was learned.	Deep reflection on how AI use connects to learning goals, skill development, and future practice.
Reproducibility	Another person could not understand or replicate the process.	Process is partially reproducible, but gaps exist.	Process is fully documented. Another person could follow the same steps.	Documentation teaches. Another person would learn to use AI effectively by reading it.

Implementation Guidance
Introducing Rubrics to Students
Share rubrics before assignments begin, not after. Students should understand how their AI use will be evaluated and can use rubrics to guide their process. Consider having students self-assess before submitting, identifying where they believe their work falls on each criterion.
Weighting Criteria
Not all criteria deserve equal weight. For any assignment, identify which criteria represent the core learning objectives and weight those more heavily. Process Documentation might be worth 10% in one assignment and 30% in another, depending on your goals.
Combining Rubrics
Many assignments will require combining elements from multiple rubrics. A research paper might use criteria from the Research Project rubric, the Writing rubric, and the Process Documentation rubric. Select the criteria most relevant to your learning objectives.
Adapting for Grade Level
These rubrics are written for high school students. For middle school, consider simplifying language and reducing the number of criteria assessed at once. For younger students, focus on a single criterion per assignment until students develop AI literacy. For older students or advanced classes, add criteria or raise expectations for 'Exemplary' performance.
The Role of Conversation
Rubrics should inform, not replace, conversation. When a student's work is difficult to assess, talk with them. Ask about their process. A student who can articulate sophisticated reasoning about their use of AI has demonstrated learning, even if the documentation is incomplete.
Evolving Standards
What counts as 'Proficient' AI use will change as both AI and students evolve. Review and update these rubrics regularly. Consider involving students in revising rubrics based on what they've learned about effective AI partnership.

* * *

The goal of these rubrics is not to police AI use but to reward thoughtful, critical, transparent engagement with AI tools. Students who learn to use AI well, to question it appropriately, and to maintain their own thinking and voice are developing skills that will serve them long after any particular AI tool is obsolete.

Appendix D: Failure Types at a Glance

Failure Type	Quick Recognition	First Response
Hallucination	Invented facts, fake citations, non-existent sources	Verify independently; name the error
Outdated Info	Information that was true but isn't anymore	Check date; use current sources
Confident Error	Wrong information is stated with certainty	Verify; discuss why confidence ≠ in truth
Math Error	Calculation mistakes, miscounting	Check with a calculator; show the work
Logic Fallacy	Flawed reasoning that sounds persuasive	Name the fallacy; reconstruct soundly
Contradiction	Saying X and not-X in the same response	Quote both; ask which is correct
Sycophancy	Agreeing too readily, changing position	Push back; test with known facts
Over-Refusal	Blocking legitimate requests	Reframe with educational context
Context Loss	Forgetting the earlier conversation	Repeat key context; start fresh
Generic Output	Bland, could-apply-to-anything content	Request specificity; ask for examples
Verbosity	Too many words, too little meaning	Ask for a concise version; specify length
Format Failure	Wrong structure for the purpose	Specify format; provide example
Bias	Reflecting societal inequities	Name it; request diverse perspectives
Cultural Assumption	Presenting particular as universal	Ask about other cultural contexts

A Final Word: Failure as Feature

Every failure documented here is something students will encounter. By explicitly addressing failures as learning opportunities rather than problems to hide, you prepare students for a world where AI is both powerful and flawed.

The student who has seen AI hallucinate will never fully trust AI output. The student who has caught AI in a math error will always verify calculations. The student who has identified AI bias will approach AI-generated content with critical awareness.

These are not unfortunate side effects of teaching with AI. They are among the most valuable lessons AI can teach.

Appendix E: Tools and Platforms to Explore

Note: This list reflects tools available as of [publication date]. AI tools evolve rapidly. Some entries may have changed, merged, or disappeared by the time you read this. For an updated list, visit [author website URL]. The underlying principles, which use AI to support learning rather than replace it, remain constant even as specific tools change.

General-Purpose AI Assistants

ChatGPT (OpenAI)

The tool that sparked the current AI-in-education conversation. A free tier is available; paid tiers offer more capabilities. Wide range of educational applications. Requires careful attention to accuracy and appropriate use policies.

chat.openai.com

Claude (Anthropic)

Alternative to ChatGPT with different strengths and limitations. Known for longer context handling and thoughtful responses. Free tier available. Worth comparing with ChatGPT for your specific use cases.

claude.ai

Gemini (Google)

Google's AI assistant is integrated with Google Workspace. It may be relevant for schools already using Google for Education—multimodal capabilities (text, image, voice).

gemini.google.com

Copilot (Microsoft)

Microsoft's AI assistant is integrated with Microsoft 365. Relevant for schools using the Microsoft ecosystem. Includes image generation and web search capabilities.

copilot.microsoft.com

Education-Specific AI Tools

Khanmigo (Khan Academy)

AI tutor built on GPT-4 but explicitly designed for educational use. Emphasizes Socratic questioning over answer-giving. Includes safeguards for student use. Subscription required.

khanacademy.org/khan-labs

Brisk Teaching

AI-powered teacher assistant for feedback, lesson planning, and content creation. Designed to save teacher time while maintaining pedagogical quality. Chrome extension integrates with standard tools.

briskteaching.com

Diffit

Creates leveled reading materials using AI. Adapts content to different reading levels while maintaining core information. Useful for differentiation.

diffit.me

Curipod

AI-powered interactive lesson creation. Generates discussion questions, polls, and activities. Integrates student response collection.

curipod.com

SchoolAI

Platform for creating AI tutors and assistants with teacher controls. Allows customization of AI behavior for specific educational contexts. Includes monitoring and safety features.

schoolai.com

Specialized Tools by Subject

Photomath

An AI-powered math problem solver that shows step-by-step solutions. Useful for student practice and understanding solution methods. Also helpful for teachers creating worked examples.

photomath.com

Grammarly

AI-powered writing assistant for grammar, style, and clarity. Now includes generative features. Consider implications for writing instruction before integrating.

grammarly.com

Quillbot

Paraphrasing and writing tool. Helpful in teaching about paraphrasing, but requires guidance on appropriate use to avoid plagiarism concerns.

quillbot.com

Canva Magic Studio

AI features are integrated into Canva's design platform. Includes image generation, writing assistance, and presentation tools. Free tier available.

canva.com/magic

AI Detection and Academic Integrity

GPTZero

An AI detection tool designed for educators. Analyzes writing for likely AI generation. Use with caution; detection is imperfect, and false positives occur.

gptzero.me

Turnitin AI Detection

Turnitin's AI detection feature is integrated with its plagiarism detection platform. Many schools already have access through existing Turnitin subscriptions.

turnitin.com

Important: AI detection tools have significant limitations. False positives can harm students. Use detection as one data point, not definitive evidence. Process-based assessment (as described throughout this book) is more reliable than detection.

References

"10 Strategies to Build on Student Collaboration in the Classroom." The Graduate School of Education and Human Development, George Washington University. https://gsehd.gwu.edu/articles/10-strategies-build-student-collaboration-classroom

"21 Language Learning Games for Your ELA Classroom." *Edutopia* (2025). https://www.edutopia.org/article/21-learning-games-ela-classroom/

Adams, Douglas. *The Hitchhiker's Guide to the Galaxy*. Pan Books, 1979.

"AI and Collaboration." *AVID Open Access*. https://avidopenaccess.org/resource/ai-and-collaboration/

"AI Delivers Creative Output but Struggles with Thinking Processes." (2025). *arXiv*. https://arxiv.org/pdf/2503.23327

"Akinator." *Wikipedia*. https://en.wikipedia.org/wiki/Akinator

Arms Control Association. (2017). "The man who 'saved the world' dies at 77." *Arms Control Today*. https://www.armscontrol.org/act/2017-10/news-briefs/man-who-saved-world-dies-77

Aura, I. et al. (2023). "Role-play experience's effect on students' 21st century skills propensity." *The Journal of Educational Research*. https://www.tandfonline.com/doi/full/10.1080/00220671.2023.2227596

"Basics of PDSA Cycles." *Carnegie Foundation for the Advancement of Teaching*. https://www.carnegiefoundation.org/improvement-products-and-services/articles/basics-of-pdsa-cycles/

Bilyk, Z.I., Shapovalov, Y.B., Shapovalov, V.B., Megalinska, A.P., Andruszkiewicz, F., Dolhańczuk-Śródka, A., & Antonenko, P.D. (2022). "Comparing Google Lens Recognition Accuracy with Other Plant Recognition Apps." *Proceedings of the 1st Symposium on Advances in Educational Technology (AET 2020)*, 2, 20-33. https://doi.org/10.5220/0010928000003364

"Biodiversity at School or in My Neighborhood Using SEEK." *National Wildlife Federation*.

https://www.nwf.org/eco-schools

Bloomberg News. (2023, June 8). "Generative AI takes stereotypes and bias from bad to worse." *Bloomberg.com.* https://www.bloomberg.com/graphics/2023-generative-ai-bias/

Boye, J., et al. (2025). "Large language models and mathematical reasoning failures." *Frontiers in Psychology.* arXiv:2502.11574

Breen, B. (2023). "Simulating History with ChatGPT." *Res Obscura.* https://resobscura.substack.com/p/simulating-history-with-chatgpt

Capps, K. (2008). "Chemistry Taboo: An Active Learning Game for the General Chemistry Classroom." *Journal of Chemical Education*, 85(4), 518.

"ChatGPT Teacher Tips Part 1: Role-Playing Activities." *EdTechTeacher.* https://edtechteacher.org/chatgptroleplaying/

Chawla, L. (2006). "Learning to Love the Natural World Enough to Protect It." *Barn*, 2, 57-78. Norwegian Centre for Child Research.

"Claude Shannon." *Wikipedia.* https://en.wikipedia.org/wiki/Claude_Shannon

"Collaborative Artificial Intelligence for Learning (CAIL)." *MIT Scheller Teacher Education Program.* https://education.mit.edu/project/collaborative-ai-for-learning-cail/

"Collaborative Learning." *Center for Teaching Innovation, Cornell University.* https://teaching.cornell.edu/teaching-resources/active-collaborative-learning/collaborative-learning

Collins, G.P. (2002). "Claude E. Shannon: Founder of Information Theory." *Scientific American.* https://www.scientificamerican.com/article/claude-e-shannon-founder/

"Considering other perspectives through role plays." *Harvard Instructional Moves.* https://instructionalmoves.gse.harvard.edu/considering-other-perspectives-through-role-plays

Davenport, M. (2016). "How to Use Socratic Seminars to Build a Culture of Student-Led Discussion." *Edutopia.*

https://www.edutopia.org/blog/socratic-seminars-culture-student-led-discussion-mary-davenport

"Debate Research." *ABLConnect, Harvard University.* https://ablconnect.harvard.edu/debate-research

Deng, Z., & Benckendorff, P. (2024). "The application and challenges of ChatGPT in educational transformation." *Heliyon.* https://pmc.ncbi.nlm.nih.gov/articles/PMC10828640/

"Descriptive Writing." *Reading Rockets.* https://www.readingrockets.org/classroom/classroom-strategies/descriptive-writing

"Design and Build a Rube Goldberg." *TeachEngineering.* https://www.teachengineering.org/activities/view/cub_simp_machines_lesson05_activity1

"Design-based learning." *Wikipedia.* https://en.wikipedia.org/wiki/Design-based_learning

Deveci, I. (2019). "Reflections of Rube Goldberg machines on the prospective science teachers' STEM awareness." *Contemporary Issues in Technology and Teacher Education*, 19(2). https://citejournal.org/volume-19/issue-2-19/science/reflections-of-rube-goldberg-machines-on-the-prospective-science-teachers-stem-awareness/

Dickerson, D.L. et al. (2013). "Effects of Modeling Instruction on Descriptive Writing and Observational Skills in Middle School." *International Journal of Science and Mathematics Education.* https://link.springer.com/article/10.1007/s10763-013-9456-2

El Majidi, A., de Graaff, R., & Janssen, D. (2021). "The effects of in-class debates on argumentation skills in second language education." *System: An International Journal of Educational Technology and Applied Linguistics*, 101. https://www.sciencedirect.com/science/article/pii/S0346251X21001305

Fan, S-C. & Yu, K-C. (2020). "Effects of infusing the engineering design process into STEM project-based learning to develop preservice technology teachers' engineering design thinking." *International Journal of STEM Education*, 8(1). https://stemeducationjournal.springeropen.com/articles/10

.1186/s40594-020-00258-9

Grüner, T., Kowalski, P., & Lewandowsky, S. (2023). "Learning about informal fallacies and the detection of fake news: An experimental intervention." *PLOS ONE.* https://pmc.ncbi.nlm.nih.gov/articles/PMC10057814/

Guilbeault, D., et al. (2025). "AI-generated faces influence gender stereotypes and racial homogenization." *Scientific Reports*, 15, 99623.

"Hallucination (artificial intelligence)." *Wikipedia.* https://en.wikipedia.org/wiki/Hallucination_(artificial_intelligence

"Harness artificial intelligence to improve classroom debates." *Times Higher Education* (2024). https://www.timeshighereducation.com/campus/harness-human-and-artificial-intelligence-improve-classroom-debates

"iNaturalist Educator's Guide." *iNaturalist Help.* https://help.inaturalist.org/en/support/solutions/articles/151000170805-inaturalist-educator-s-guide

"Information Theory." *CS Unplugged.* https://classic.csunplugged.org/activities/information-theory/

Jiang, C. & Pang, Y. (2023). "Enhancing design thinking in engineering students with project-based learning." *Computer Applications in Engineering Education.* https://onlinelibrary.wiley.com/doi/abs/10.1002/cae.22608

Kara, K. & Yildirim, K. (2021). "The Views of the Seventh Grade Students on Learning English Vocabulary with the Taboo Game." *International Social Sciences Education Journal*, 7(1). https://dergipark.org.tr/en/pub/issej/issue/60224/910488

Koivisto, M. & Grassini, S. (2023). "Best humans still outperform artificial intelligence in a creative divergent thinking task." *Scientific Reports*, 13, 13601. https://www.nature.com/articles/s41598-023-40858-3

Kuo, M., Barnes, M., & Jordan, C. (2019). "Do Experiences With Nature Promote Learning? Converging Evidence of a Cause-and-Effect Relationship." *Frontiers in Psychology*, 10, 305.

https://pmc.ncbi.nlm.nih.gov/articles/PMC6401598/

Lane, J.N., Bouissioux, L., Zhang, M., Lakhani, K. & Jacimovic, V. (2024). "Can AI Match Human Ingenuity in Creative Problem-Solving?" *Harvard Business School Working Knowledge.*
https://www.library.hbs.edu/working-knowledge/generative-ai-and-creative-problem-solving

Language Log. (2023). "ChatGPT writes haiku." *University of Pennsylvania.*
https://languagelog.ldc.upenn.edu/nll/?p=57468

Lumbangaol, R.R. (2021). "The Effect of Taboo Word Game in Improving Vocabulary Ability." *ResearchGate.*
https://www.researchgate.net/publication/361292553

MacDonald, J. (2024). "Dude, where's my citations? ChatGPT's hallucination of citations." *Mind Pad*, Winter 2024. Canadian Psychological Association.

Mann, J., Gray, T., Truong, S., Brymer, E., Passy, R., Ho, S., Sahlberg, P., Ward, K., Bentsen, P., Curry, C., & Cowper, R. (2022). "Getting Out of the Classroom and Into Nature: A Systematic Review of Nature-Specific Outdoor Learning on School Children's Learning and Development." *Frontiers in Public Health*, 10, 877058.
https://www.frontiersin.org/articles/10.3389/fpubh.2022.877058/full

Marks, J. & Chase, C.C. (2019). "Impact of a prototyping intervention on middle school students' iterative practices and reactions to failure." *Journal of Engineering Education*, 108(4), 547-573.
https://onlinelibrary.wiley.com/doi/abs/10.1002/jee.20294

Mata v. Avianca, Inc., 678 F.Supp.3d 443 (S.D.N.Y. 2023).
https://en.wikipedia.org/wiki/Mata_v._Avianca,_Inc.

McGrew, S. et al. (2018). "Improving University Students' Web Savvy: An Intervention Study." *British Journal of Educational Psychology.* Stanford History Education Group.

Meincke, L. et al. (2025). "Does AI Limit Our Creativity?" *Knowledge at Wharton.*
https://knowledge.wharton.upenn.edu/article/does-ai-limit-our-creativity/

Mitra, S. (2012). "The Hole in the Wall Project and the Power of

Self-Organized Learning." *Edutopia.* https://www.edutopia.org/blog/self-organized-learning-sugata-mitra

Mosier, K. L., Skitka, L. J., Heers, S., & Burdick, M. D. (1998). "Automation bias: Decision making and performance in high-tech cockpits." *International Journal of Aviation Psychology*, 8(1), 47-63.

Nemri, I. (2024). "How Akinator Reads Your Mind: Unveiling the Game's Algorithmic Magic." *Medium.* https://medium.com/@inemri/how-akinator-reads-your-mind-unveiling-the-games-algorithmic-magic-c8ee86dbc1d3

News Literacy Project. "Fact-check it!" https://newslit.org/educators/resources/fact-check-it/

npj Digital Medicine. (2025). "When helpfulness backfires: LLMs and the risk of false medical information due to sycophantic behavior."

Ortiz, G. (2025). "AI has become the norm for students. Teachers are playing catch-up." *NBC News.* https://www.nbcnews.com/tech/tech-news/ai-school-teacher-student-train-chatgpt-rcna248726

"Outdoor Education Research Summary." *Wisconsin K-12 Forestry Education Program (LEAF).* University of Wisconsin-Stevens Point. https://www.uwsp.edu/cnr-ap/leaf/

Paul, R. & Elder, L. (2007). "Critical Thinking: The Art of Socratic Questioning." *Journal of Developmental Education*, 31(1), 34-37.

"PDSA Cycle." *The W. Edwards Deming Institute.* https://deming.org/explore/pdsa/

Poynter Institute. (2019). "Want to be a better fact-checker? Play a game." https://www.poynter.org/fact-checking/2019/want-to-be-a-better-fact-checker-play-a-game/

Roozenbeek, J. & van der Linden, S. (2019). "Fake news game confers psychological resistance against online misinformation." *Palgrave Communications.*

Rozear, H. & Park, S. (2023). "ChatGPT and Fake Citations." *Duke University Libraries Blogs.* https://blogs.library.duke.edu/blog/2023/03/09/chatgpt-

and-fake-citations/

Scager, K. et al. (2016). "Collaborative Learning in Higher Education: Evoking Positive Interdependence." *CBE Life Sciences Education*, 15(4). https://www.lifescied.org/doi/10.1187/cbe.16-07-0219

Schueler, B.E. & Larned, K.E. (2024). "Interscholastic Policy Debate Promotes Critical Thinking and College-Going: Evidence From Boston Public Schools." *American Educational Research Journal*. https://journals.sagepub.com/doi/10.3102/01623737231200234

"Seek by iNaturalist." *iNaturalist*. https://www.inaturalist.org/pages/seek_app

"Seek by iNaturalist Review for Teachers." *Common Sense Education*. https://www.commonsense.org/education/reviews/seek-by-inaturalist

Shannon, C.E. (1948). "A Mathematical Theory of Communication." *Bell System Technical Journal*, 27(3), 379-423. https://people.math.harvard.edu/~ctm/home/text/others/shannon/entropy/entropy.pdf

Shapovalov, Y.B., Shapovalov, V.B., Bilyk, Z.I., Megalinska, A.P., & Antonenko, P.D. (2019). "The Google Lens Analyzing Quality: An Analysis of the Possibility to Use in the Educational Process." *CEUR Workshop Proceedings*. https://www.researchgate.net/publication/340050262

Sharma, M., et al. (2023). "Towards understanding sycophancy in language models." *arXiv preprint*, arXiv:2310.13548.

Siregar, M.A. & Fithriani, R. (2023). "Learning English Vocabulary with Taboo Game: A Case Study of Indonesian Junior High School Students." *Ethical Lingua: Journal of Language Teaching and Literature*, 8(1). https://ethicallingua.org/25409190/article/view/513

"Socratic Questioning." *Wikipedia*. https://en.wikipedia.org/wiki/Socratic_questioning

"SOCRATIC SEMINARS IN SCIENCE CLASS: Providing a structured format to promote dialogue and understanding." *The American Biology Teacher*.

https://pmc.ncbi.nlm.nih.gov/articles/PMC4322762/

"Socratic Seminars." *ReadWriteThink*, National Council of Teachers of English. https://www.readwritethink.org/professional-development/strategy-guides/socratic-seminars

Stanford History Education Group. (2016). "Evaluating Information: The Cornerstone of Civic Online Reasoning." *Stanford Graduate School of Education*. https://ed.stanford.edu/news/stanford-researchers-find-students-have-trouble-judging-credibility-information-online

Stanford History Education Group. (2019). "Students' Civic Online Reasoning." *Digital Inquiry Group*. https://www.inquirygroup.org/students-civic-online-reasoning

"Students' Perceptions on the Effectiveness of the Taboo Game in Enhancing English Vocabulary Acquisition." *Elsya: Journal of English Language Studies* (2024). https://journal.unilak.ac.id/index.php/elsya/article/view/19819

Sun, L., Wei, M., Sun, Y., Suh, Y. J., Shen, L., & Yang, S. (2024). "Smiling women pitching down: Auditing representational and presentational gender biases in image-generative AI." *Journal of Computer-Mediated Communication*, 29(1), zmad045.

"Supporting PBL with a Design Thinking Framework." *PBLWorks*. https://www.pblworks.org/blog/supporting-pbl-design-thinking-framework

"Teaching and Learning Vocabulary with AI." *Hong Kong TESOL*. https://hongkongtesol.com/blog/teaching-and-learning-vocabulary-ai

"The Power of Speech & Debate Education." *Stanford National Forensic Institute*. https://snfi.stanford.edu/skills

"The Socratic Method: Fostering Critical Thinking." *The Institute for Learning and Teaching, Colorado State University*. https://tilt.colostate.edu/the-socratic-method/

"The Use of Socratic Questioning in Science Inquiry-Based Learning." *SingTeach*. https://singteach.nie.edu/2021/09/16/the-use-of-socratic-questioning-in-science-inquiry-based-learning/

Tse, D. (2020). "How Claude Shannon Invented the Future." *Quanta Magazine*. https://www.quantamagazine.org/how-claude-shannons-information-theory-invented-the-future-20201222/

"University Study Hints Debate Skills May Lessen Negative Impact of AI." *University of Mississippi News* (2024). https://olemiss.edu/news/2024/09/debate-ai-debate/index.html

"Using PDSA Cycles to Boost Learning Outcomes." *Edutopia* (2023). https://www.edutopia.org/article/using-pdsa-cycles-boost-learning-outcomes

Walters, W.H., & Wilder, E.I. (2023). "Fabrication and errors in the bibliographic citations generated by ChatGPT." *Scientific Reports*, 13, 14045. https://www.nature.com/articles/s41598-023-41032-5

Wang, D. et al. (2025). "The effects of generative AI on collaborative problem-solving and team creativity performance in digital story creation." *International Journal of Educational Technology in Higher Education*, 22. https://educationaltechnologyjournal.springeropen.com/articles/10.1186/s41239-025-00526-0

Wang, P., et al. (2025). "One year in the classroom with ChatGPT: Empirical insights and transformative impacts." *Frontiers in Education*. https://www.frontiersin.org/journals/education/articles/10.3389/feduc.2025.1574477/full

Zandvoort, H. et al. (2013). "Using and Developing Role Plays in Teaching Aimed at Preparing for Social Responsibility." *Science and Engineering Ethics*, 19(4). https://ncbi.nlm.nih.gov/pmc/articles/PMC3857546

Zheng, L., Niu, J., & Zhong, L. (2023). "Effects of chatbot-assisted in-class debates on students' argumentation skills and task motivation." *Computers & Education*, 203, 104862. https://www.sciencedirect.com/science/article/abs/pii/S0360131523001392

www.ingramcontent.com/pod-product-compliance
Lightning Source LLC
LaVergne TN
LVHW020933090426
835512LV00020B/3335